Tomas Magalhaes
Pedro Brás de Oliveira
Cristina Oliveira

Propagation and phenology of Corema album (L.) D.Don

Tomas Magalhaes
Pedro Brás de Oliveira
Cristina Oliveira

Propagation and phenology of Corema album (L.) D.Don

Propagation by cuttings and proposal for a BBCH scale

Imprint

Any brand names and product names mentioned in this book are subject to trademark, brand or patent protection and are trademarks or registered trademarks of their respective holders. The use of brand names, product names, common names, trade names, product descriptions etc. even without a particular marking in this work is in no way to be construed to mean that such names may be regarded as unrestricted in respect of trademark and brand protection legislation and could thus be used by anyone.

Cover image: www.ingimage.com

This book is a translation from the original published under ISBN 978-3-330-77381-3.

Publisher:
Sciencia Scripts
is a trademark of
Dodo Books Indian Ocean Ltd. and OmniScriptum S.R.L publishing group

120 High Road, East Finchley, London, N2 9ED, United Kingdom
Str. Armeneasca 28/1, office 1, Chisinau MD-2012, Republic of Moldova, Europe
Managing Directors: Ieva Konstantinova, Victoria Ursu
info@omniscriptum.com

Printed at: see last page
ISBN: 978-620-3-25436-5

Index

Propagation and phenology of *Corema album* (L.) D.Don

Thanks

To Researcher Pedro Bras de Oliveira of Linking Landscape, Environment, Agriculture and Food (LEAF), Instituto Nacional de Investigagao Agraria e Veterinaria, I. P. (INIAV), for his support, knowledge, objectivity and availability. (INIAV), for his support, knowledge, objectivity, availability and help. To Professor Cristina M. M. Simoes Oliveira from Linking Landscape, Environment, Agriculture and Food (LEAF), Instituto Superior de Agronomia (ISA), University of Lisbon, for her motivation, support, objectivity, availability and help.

To the researcher Teresa Valdiviesso for her great help with phenology, Portuguese and formatting.

To Francisco Luz, Francisco Barreto, Xana, Francisca, Fernanda, Càndida, and all the INIAV researchers and staff who helped me, even when it was dark in the greenhouse.

To my parents and their partners, for their enormous support, motivation and interest, for giving me the happiness they taught me to feel.

To my partner for simplifying everything, for making me feel lucky and for everything.

To my sister and the Magalhaes Brito family, who always make me proud and inspire me.

To my brothers and modern family for the moments of sharing and always ready help.

To class 1, to all the influential characters in this class for the great moments and friendships.

To INIAV for allowing me to use their facilities.

To all my friends and family for the inspiration, the moments and the friendships.

To my Kangoo who has done so many kilometers without complaining too much. Thanks again to my parents, because without their support there would be no Kangoo to thank, and she would certainly complain a lot more.

To Novas Ediçoes Académicas for publishing this work

Summary

The dioecious species *Corema album*, endemic to the Atlantic coast of the Iberian Peninsula, has the potential to become part of the small fruit market. Trials were carried out on vegetative propagation and phenological characterization, proposing a BBCH scale for this species. Survival and rooting were compared between: 1) two genotypes (wild and cultivated) treated with auxin (0, 500, 1000 and 1500 ppm); 2) nine origins in two substrates ("Siro" and "Fataca"). Genotype influenced survival and rooting. After 165 days, the "Aldeia do Meco Selvagem" genotype showed the best rooting (59.3%) and "Aldeia do Meco Cultivado" the worst survival (40.0%). Auxin had no significant effect. Substrate and place of origin influenced survival and rooting. The Siro substrate showed the best survival (77.4%) and rooting (74.6%). The best origin was Vila Real de Santo António with 63.6% rooted. The phenological observations took place between January and September in Aldeia do Meco. The vegetative growth of the male and female plants is similar, with temporal differences. The plants flowercd synchronously and the fruit formed and developed between March 27 and August 16. As it is a dioecious plant with two types of vegetative growth, 3 subphases were added to the BBCH scale, resulting in 7 phases and 3 subphases.

Key words: *Corema album*, vegetative propagation, phenological characterization, small fruits, BBCH

1. introduction

The camarinha (*Corema album* (L.) D.Don 1830) is a dioecious shrub endemic to dune systems on the Atlantic coast of the Iberian Peninsula, belonging to the *Ericacea* family (Oliveira and Dale, 2012). This shrub produces a white drupe that has a citrus flavor and contains a high amount of antioxidants (Santos, *et al.* 2009). The white drupe can be included in the category of small fruits, being a unique small fruit due to its white color and very different taste. The fact that it is a very particular fruit, with interesting qualitative and taste characteristics, means that it has enormous potential for introduction into the small fruit market.

There are few studies on the species *Corema album* (L.) D.Don 1830, and it is necessary to investigate various agronomic aspects of this species in order to broaden the range of existing information and consequent knowledge of the species.

In this work, two areas were analyzed, propagation and phenology, because they are areas that, in addition to combining, provide key information for making decisions in terms of production, as well as helping to analyze the next steps to be taken in the area of research.

Vegetative propagation (rooting by cuttings) was addressed, as no published work was found on this species. In the area of sexual propagation, Santos (2013) and Santos *et al.* (2014) carried out different pre-treatments on seeds from different locations in Portugal, and observed an influence of the different locations and pre-treatments. Given the influence of location on seed propagation, it was decided to carry out the study with the variable of the origin of the plant material, to see if this variability is maintained in the case of vegetative propagation.

As a basis for decisions such as: the type of cutting to choose; rooting methods; treatments to apply and, in particular, the application of phyto-regulators for rooting, we used the work of various species of the Ericacea family (Bowerman *et al.*, 2012; Mckechnie *et al.*, 2012; Magnitskiy *et al.*, 2011; Mihaljevic and Salopek-Sondi, 2012; Celik and Obadas, 2009; Maragon and Biasi, 2013 and Vignolo *et al.*, 2012).

The choice of substrate is very important in the success of propagation by cuttings, and according to Macdonald (1983), in addition to the different intrinsic characteristics of the substrate, the species to be propagated has a major influence on the choice. In the works by Tchoundjeu *et al.* (2002), Mesén *et al.* (1996) and Klein *et al.* (2000), some differences in rooting success were observed when comparing different substrates.

In the area of vegetative propagation of camarinha, a first objective was to compare the rooting capacity of wild and cultivated genotypes and to observe their response to different concentrations of auxin. A second objective was to compare the rooting capacity of different genotypes from Portugal and to check the suitability of two substrates for this stage of plant development.

With regard to the phenology of camarinha, the work by Alvarez-Casino *et al.* (2010) focuses a lot on the influence of the different sexes on vegetative growth. This work covers the subject of phenology but does not describe the phenology of the species. In the case of the work by Oliveira and Dale (2012), there is a general characterization of the camarinha, in which a paragraph is dedicated to phenological aspects. Guitian *et al.* (1997) and Zunzunegui *et al.* (2006) describe various aspects of the camarinha's phenology. Based on these works, a phenological characterization of the species was carried out, checking what coincides with what has been described previously, introducing a phenological scale with the aim of schematizing and organizing the different phenological phases. To do this, we used the BBCH scale described in Meier *et al.* (2008), as it is the most widely accepted scale in the scientific community and standardizes phenology quite efficiently for all species.

In conjunction with the two trials and the phenological characterization, several questions arise: 1) Do the origin and physiological characteristics influence the survival and/or rooting of the cuttings; 2) Do the different substrates influence the survival and/or rooting of the cuttings;

3) Do the different auxin concentrations have an effect on the survival and/or rooting of the cuttings, and is this effect positive or negative; 4) Which phenological phases coincide with each other, and how do they fit into time?

5) Are there differences between types of shrubs in terms of phenology and the timing of the different stages? 6) Does the BBCH scale serve to describe the totality of phenological events? 7) At which phenological stages would cuttings be taken with the greatest chance of survival/rooting?

This work aims to provide a contribution and answers to these questions, enriching our knowledge of the physiology of this peculiar species.

2. Bibliographical review

2.1 Species *Corema album*

Part of the Ericaceae family, sub-family Ericoideae, the species *Corema album* (L.) D.Don is one of two species in the *Corema* genus, the other being *Corema conradii* (Torr.) Torr. ex Loudon (Oliveira and Dale, 2012).

Corema conradii is found in eastern North America. This species is distinguishable from *C.album* because it has much smaller fruits, which lack fleshy parts and are covered with oily appendages (Martine *et al.*, 2005).

According to Valdés *et al.* (1987) there are two subspecies of *C.album*, subsp. *album* and subsp. *azoricum* P.Silva. The subsp. *album* occurs on the Atlantic coast of the Iberian Peninsula, from Gibraltar to Finisterre and the subsp. *azoricum*, as the name suggests, occurs in the Azores (Oliveira and Dale, 2012).

Corema album occurs mainly in dunes, but can also occur in rocky areas and volcanic fields, as in the Azores (Calvino-Cancela, 2004).

It is a dioecious shrub with many branches and a normal height of 30 to 75 cm, which can reach up to 1 m. The leaves are in whorls of 3 or 4, with short petioles that tend to lean against the branch, and are 8 to 10 mm long and 1 mm wide (Oliveira and Dale, 2012).

Corema album is a dioecious species, however 1 to 4 % of male plants in the southeast (Vila Real de Santo Antònio and Donana) have some hermaphrodite inflorescences (Zunzunegui *et al.*, 2005).

The flowers are actinomorphic (they can be divided into 3 or more identical radially symmetrical sectors) and appear in terminal inflorescences in groups of 4 to 14. The male flowers have three 2 to 3 mm suborbicular and pubescent sepals, three 3 to 5 mm pink petals and three 5 to 6 mm stamens with very visible red anthers. The female flowers are smaller, have sepals of 1 to 2 mm, and petals of 1 mm, the ovary is superior with a stipe present, and the stigma has three red lobes (Simonds, 1979 and Tutin *et al.*, 1972).

The fruits are white or pinkish white and are berry-like drupes 5 to 8 mm in diameter, usually with three pyrenes (seeds) (Simonds, 1979 and Tutin *et al.*, 1972). The fruit is edible, with a pleasant and very refreshing taste (Gonzalez, 2001).

Ripe fruits can be pearly white, others can be more translucent (you can see the seeds), and sometimes they can be reddish or greenish in color (Oliveira and Dale, 2012).

The berries have a moderately acidic and citric flavor. They contain many antioxidants, low amounts of anthocyanins, high amounts of flavinol, chlorogenic acid derivatives and phenolic acid (Santos *et al.*, 2009).

The fruit has been eaten fresh for centuries and is sold in some public markets in Galicia. It is used in traditional medicine to reduce fevers and kill roundworms (Gonzalez, 2001).

2.2 Propagation techniques

There are two main branches of plant propagation, one being seed (or sexual) propagation and the other vegetative (or asexual) propagation, each of which has its advantages and disadvantages. This chapter focuses on vegetative propagation, but the advantages and disadvantages are described in comparison with seed propagation.

A very important feature, which can be seen as an advantage, of vegetative propagation is the fact that you get plants that contain the same characteristics as the plant that was propagated (mother plant).

According to Macdonald (1993), some of the advantages of vegetative propagation over propagation by seed are: achieving high crop uniformity, overcoming the problems of dormancy (which, according to Santos (2013), is high in the *Corema album* species) and low seed viability, reducing the time needed to start production and avoiding the transmission of viruses via seed.

With regard to the disadvantages of vegetative propagation compared to seed propagation, according to Macdonald (1993): viruses can be transmitted via vegetative propagation, disease resistance is not increased, lower vigor (influence of the juvenile phase), usually higher cost, low productivity.

According to Macdonald (1993), the main methods of vegetative propagation are: propagation by cuttings, grafting (fork, bud, among others), micropropagation, divisions and divisions. Deciding which technique to use and which part (and size) of the plant to use depends on various factors such as: the species to be propagated, the equipment/technology and sites available and the skills/experience of the propagator.

The choice of substrate to use is very important not only for rooting but also for the survival of the plant material placed in it. There are various criteria to be analyzed when choosing the mixture of materials present in the substrate, which according to Macdonald (1993) are cost (transport and material itself), quality (pH, particle size, impurities, among others) and structure (support, aeration and water retention) and also the species to be propagated and its specific characteristics. In the case of Ericacea, the most favorable pH for rooting is 4.0 to 5.0 (Macdonald, 1993).

The type of substrate used influences rooting, for example according to Tchoundjeu *et al.* (2002) there was a difference of 9% in rooting and 15% in mortality of plants of the species *Prunus africana* between two types of substrate, and according to Mesén *et al.* (1996) two substrates sand and gravel had significantly higher rooting percentages than the substrate with sawdust in cuttings of *Cordia alliodora*. According to Klein *et al.* (2000) who used two similar substrates, one with 30% more perlite than the other, only one variety of *Myrtus communis* had significantly different responses to the two substrates.

2.2.1 The formation of roots

Vegetative propagation is based on the formation of roots using different plant organs, often from the aerial part of the plant. This root formation is complex and requires specific conditions to take place.

There are three types of root: the main root, the lateral roots and the adventitious roots. All plants have a main root (which evolved from the embryonic root), and different types of lateral roots. According to Bellini *et al.* (2014), adventitious roots perform the same functions as lateral roots, but are formed from tissues of aerial organs. However, some authors consider that the difference between lateral and adventitious roots is anatomical: while lateral roots originate in the cells of the pericycle, adventitious roots are formed in stems, leaves and tissues other than the pericycle in older roots (Li *et al.*, 2009). It seems that lateral and adventitious roots share the same key elements of hormonal and genetic regulation, but do so through different mechanisms. Although more and more is known about the development, regulation and environmental adaptation of adventitious roots, this knowledge is still not as comprehensive as that of the development of main and lateral roots (Bellini *et al.*, 2014).

2.2.2 The formation of adventitious roots

Adventitious rooting is one of the most important methods of vegetative propagation and one of the most important methods of commercial production of horticultural crops worldwide (Li *et al.*, 2009).

Adventitious roots have two rooting patterns: direct and indirect. In the direct pattern, the tissues involved in rooting are the vascular tissues and those of the vascular cambium, which undergo the first mitotic divisions giving rise to the root primordia; in the indirect pattern, *callus* formation precedes the root primordia (Altamura, 1996).

These two patterns of adventitious root development have in common the three phases of root formation: induction, initiation and expression. Induction requires a higher concentration of auxin, initiation and expression are inhibited by high concentrations of auxin and are the phases in which the anatomical differentiation of the roots occurs (Costa *et al.*, 2013).

The induction phase is usually the result of wounds or cuts in the plant. At this stage there is an increase in the concentration of phenolic components, jasmonate and auxins at the base of the cutting (cut or wound zone), associated with a decrease in peroxidase activity. This base zone also becomes a carbohydrate sink zone (Schwambach *et al.* 2008; Ahkami *et al.*, 2009).

In the initiation phase, cell division occurs, meristemoids and root primordia develop. The expression phase sees the growth of root primordia and the establishment of vascular connections between the newly formed roots and the cutting (Costa *et al.*, 2013).

2.2.3 The influence of environmental factors on the development of adventitious and collateral roots

The mineral nutrition of the mother plant is very important in the capacity for adventitious rooting, intervening in the structure of the various components linked to the transport and reception of auxins (zinc), and of the peroxidases (manganese and iron) catabolizing the effect of auxin. High concentrations of nitrogen in herbaceous cuttings seem to promote adventitious rooting (Druege *et al.*, 2000, 2004; Zerche and Druege, 2009).

Successful rooting depends very much on the responsiveness of the cells. Cells from more recent tissues have a greater capacity for differentiation (Costa *et al.*, 2013).

Carbohydrates contribute to the formation of adventitious roots by providing the energy and carbons needed: for ceiuiar division, for the establishment of new root meristems and for the formation of the root itself. The transportation of these carbohydrates to the new sink zone (the cutting zone) is crucial (Costa *et al.*, 2013).

In addition to the mineral nutrition of the plant and its ability to form roots, it is necessary to place the organs to be rooted in environments that are conducive to rooting. The time of cutting and immediately after cutting is very important, as this is when the cuttings respond to the cut.

Cuttings need to maintain a positive water balance in order to form roots. Water-stressed cuttings root less and more slowly, and the percentage of survival is lower (Puri and Thompson, 2003).

High concentrations of auxin shortly after cutting can be beneficial for rooting, and little light in the cutting zone induces endogenous auxin production (Costa *et al.*, 2013). The exogenous application serves to assist the endogenous production of the same hormone, helping the rooting response. Most of the auxin applied acts in the contact zone (cutting zone of the cutting), while another part is transported along the xylem (Osterc and Spethmann, 2001 cited in Costa *et al.*, 2013).

2.2.4 The role of phytoregulators

Adventitious rooting can be compared to a stress-induced reprogramming of branch cells (Costa *et al.*, 2013), and there are various compounds, molecules and phytoregulators that take part in this reprogramming.

Among the endogenous factors that influence root formation, the phytoregulators are by far the most important. Figure 1 summarizes the interactions between phytoregulators and between these and the environment, with auxin playing an important role, especially in the early stages of root production (Bellini *et al.*, 2014).

Phenolic compounds are known to protect plants from oxidative stress, can protect auxins (preventing them from oxidizing) and improve rooting (De Klerk *et al*, 2011). We therefore want a moderate response to the cutting, enough to increase the concentration of auxins, but not too much as to oxidize the cutting or the components needed for rooting.

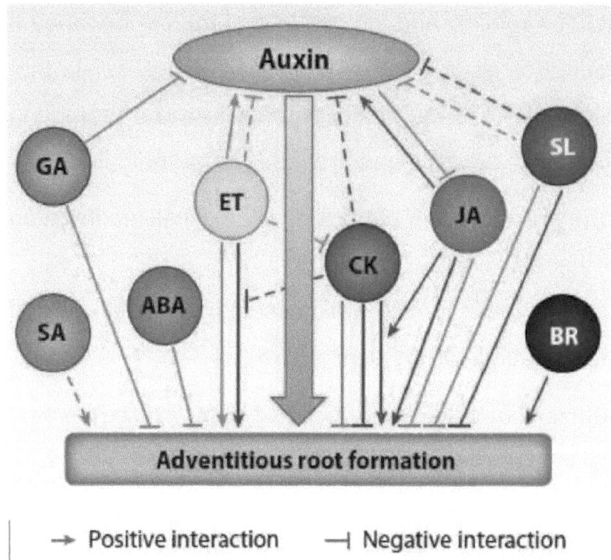

Figure 1: Cross-regulatory hormonal interactions present in the regulation of adventitious root formation in various dicotyledonous species. Several phytoregulators, including ethylene (ET), cytokinins (CK), gibberellic acid (GA), abscisic acid (ABA), strigolactones (SL), brassinosteroids (BR), salicylic acid (SA), jasmonic acid (JA), directly influence root development by interacting with each other or by interacting with auxin. The discontinuous lines represent possible links that have not yet been fully demonstrated. The orange lines represent hormonal interactions during the formation of adventitious roots in intact plants, and the blue lines represent interactions in cuttings and alternative rooting systems (adapted from Bellini *et al*, 2014).

Among the various phyto-regulators involved in root formation, jasmonic acid seems to be the one that initiates the induction of rhizogenesis. Auxins have a rhizogenic action in the induction phase of root formation (generally from the cut to 96 h after the cut) and stimulate cells in the cut to create new meristems (Garrido *et al.*, 2002), the same phytoregulators become inhibitors after 96 h and can inhibit or even block root growth and emergence. Auxins are degraded by peroxidases (Costa *et al.*, 2013).

Ethylene promotes adventitious rooting and inhibits lateral rooting because it affects auxin transport in different ways (Negi *et al.*, 2010). It is possible that gibberellins have an inhibitory effect in the induction phase, and a promoting effect in more advanced phases of adventitious rooting (Costa *et al.*, 2013).

Cytokinins and auxins are the main phytoregulators involved in cell proliferation and are indispensable in the cell cycle (Dewitte and Murray, 2003). Plant development depends on meristematic growth, which occurs when cell division takes precedence over differentiation (Costa *et al.*, 2013).

The size of the root meristem is controlled by the balance between cell division and differentiation, where auxins and cytokinins act as antagonists and play a central role (Dello Ioio *et al.*, 2007, 2008; Moubayidin *et al.*, 2009, 2010). The apical root meristem is made up of sets of cells that renew themselves and are not differentiated. There is a center called the quiescent (or resting) center that maintains these characteristics in the cells surrounding the meristem (Van den Berg *et al.*, 1997; Osmont *et al.*, 2007; Arnaud *et al.*, 2010).

The quiescent center has a low mitosis rate and has cells that are histologically different from the surrounding cells (Doener, 1998). This center functions as a reservoir of cells for regeneration to ensure the persistence of the meristematic apex. Hormonal activity is very important in the maintenance and organization of the quiescent center (Sabanti *et al.*, 1999; Ortega-Martinez *et al.*, 2007), auxins promote the organization and maintenance of this center (Sabanti *et al.*, 1999), but auxin is not enough, ethylene is needed (Ortega-Martinez *et al.*, 2007).

Recent studies suggest that stress-related phytoregulators and DNA damage can initiate cell division in the quiescent center, resulting in the replenishment of cells around the quiescent center (Heyman *et al.*, 2014). These authors suggest that the quiescent center represents a pool of backup cells that serve to replace damaged cells, contributing to the longevity of the plant.

Auxins are decisive for the correct formation of the plant's vertical axis and are transported in two ways: the fast (10 cm per hour) or apolar transport, which is bidirectional and occurs in the phloem sap, and the slow or polar transport mediated by transporters (Kerr and Bennett, 2007). There are more studies with the auxin IAA (indole-3-acetic acid) showing that both transport systems are necessary in rhizogenesis.

In polar transport, the auxin AIA (indole-3-acetic acid) transporters are specific and allow intercellular transport in branches, the transport is active polar and basipetal. AIA can be found in anionic and protonic forms, the latter being more lipophilic and capable of diffusing through the plasma membrane (Woodward and Bartel, 2005; Zazimalova *et al.*, 2010), while the anionic forms do not have the capacity to penetrate cells and therefore auxin transporter proteins are needed, which are permease-type amino acid proteins of the AUX1/LAX family (reviewed in Vieten *et al.*, 2007). These proteins function as H+/AIA-transporters and may participate in the emergence of lateral roots and the development of root hairs (reviewed in Vanneste and Friml, 2009). PIN proteins (secondary transporters) are involved in the flow of auxins and their asymmetric distribution in the cells is essential for polar basipetal transport in the branches.

Considering other auxins such as the endogenous AIB (indolbutyric acid) and synthetic ones such as ANA (1-naphthylacetic acid), relatively little is known about their transport and metabolism. AIB is more stable than AIA and remains in the plant tissues for longer (De Klerk *et al.*, 1999), and appears to have different transporters than AIA (Strader and Bartel, 2011). ANA is more stable than the above and is probably also transported by different agents (Yang *et al.*, 2006).

The formation of auxin gradients, originated by processes of biosynthesis, conjugation and degradation, and by inter- and intra-cellular transport, is relevant to morphogenesis and the determination of patterns in plant tissues (Vanneste and Friml, 2009; Overvoorde *et al.*, 2010; Simon and Petrasek,

2011). Previous studies indicate that PIN proteins play a key role in the directional distribution networks that control auxin maxima and gradients in the course of various developmental processes (reviewed in Vieten *et al.*, 2007), these auxin gradients are important in root organogenesis, both for the primary root and for lateral roots.

However, relatively little is known about the effect of polar and non-polar auxin transport on adventitious rooting. Some studies in which polar transport was inhibited show that this type of transport is important in the formation of adventitious roots (Nordstrom and Eliasson, 1991 cited in Costa *et al.*, 2013; Liu and Reid, 1992 cited in Costa *et al.*, 2013; Koukourikou Petridou and Bangerth, 1997; Guerrero *et al.*, 1999; Garrido *et al.*, 2002; Nicolas *et al.*, 2004). Considering cuttings for vegetative propagation, progressive accumulation and local concentration seem to be important to originate the peak necessary for the beginning of the rooting process (Acosta *et al.*, 2009) and can very commonly be facilitated by adding auxins in the case of recalcitrant species.

Auxins act through selective proteolysis and weaken the cell wall, but as much as it is known that auxins play a key role in adventitious rooting, the specific mechanisms of auxin action are far from understood. At the cellular level, auxins induce rapid transformations in the cell's physiology, such as membrane depolarization, acidification of apoplasts, weakening of cell membranes, activation of plasma membrane ATPases and gene control (Scherer, 2011).

Adventitious root primordia with an apical meristem and differentiation of the basic body of the root are formed and grow through the cortex and towards the surface of the stem. Ethylene seems to be important in inducing the weakening of the cell membrane and facilitating the passage of the root through stem tissues (Vidoz *et al.,* 2010).

When the newly formed roots reach the surface of the stem, the epidermis is disrupted and the cell wall is further weakened, leading to root emergence, the stem then develops a periderm wrapped around the opening of each adventitious root formed to protect it from external agents and dryness, the vascular reconnection between the new roots and the stem is fully re-established, enabling the root to nourish, hydrate and grow (Hatzilazarou *et al... 2006*), 2006). In the process of vascularization and vascular connection, auxins and cytokinins are relevant to the differentiation of phloem and xylem.

2.2.5 Propagation of shrimp and other species of the Ericacea family

The *C. album* species can be propagated via seminal propagation with pre-treatment, as demonstrated by Costa (2011) and Santos *et al.* (2014) or via vegetative propagation through cuttings. Although there are few publications on the propagation of *C. album*, and in particular vegetative propagation, there are studies on the vegetative propagation of species from the *Ericaceae* family.

Many endogenous and environmental factors, such as temperature, light, phytoregulators (especially auxin), sugars, mineral salts and other molecules, can act as signals and induce groups of cells to differentiate, acting as regulators of adventitious rooting (Li, 2009, Costa *et al.*, 2013). The success of propagation is also affected by the genome, physiological condition, age, and health status of the mother plant as well as the propagation technique and environmental rooting conditions.

The choice of the type of cuttings influences their survival and rooting percentage. In *Vaccinium arboreum*, herbaceous cuttings rooted better than woody and semi-woody cuttings (Bowerman *et al.*, 2012). In *Vaccinium membranaceum* and *Vaccinium myrtilloides*, woody *cuttings* rooted less well than semi-woody *cuttings* (Mckechnie *et al.*, 2012). In *Vaccinium floribundum* and *Disterigma alaternoides*, herbaceous cuttings rooted more than semi-woody ones (Magnitskiy *et al.*, 2011). These studies highlight the ability of herbaceous cuttings in the *Ericaceae* to form adventitious roots.

There are different auxins used to increase the rooting percentage, such as AIB (indolbutyric acid), AIA (indole-3-acetic acid) and ANA (1-naphthylacetic acid). For different species, the different phyto-regulators have different results, and the concentrations used influence the final result.

Table 1 and 2 show the results of published studies on the rooting of cuttings in species of the *Ericaceae* family.

Table 1: Effect of auxins on different species of the *Ericaceae* family, table drawn up according to the authors referenced.

Species	AIB (ppm)				ANA (ppm)		References
	200 a 1 000	1 000 a 5000	5 000 a 10 000	10 000 a 20 000	200 a 400	20 000	
Vaccinium arboreum (10s)	n	0	0	n	n	n	Bowerman *et al.* (2012)
Vaccinium membranaceum	n	n	n	+	n	0	Mckechnie *et al.* (2012)
Vaccinium myrtilloides	n	n	n	+	n	0	
Vaccinium ashei (10s)	n	+	++	n	n	n	Vignolo *et al.* (2012)
Vaccinium ashei (10s)	+	++	n	n	n	n	Maragon and Biasi (2013)
Vaccinium corymbosum (2 min)	+	++	n	n	n	n	Celik and Obadas, (2009)
Vaccinium corymbosum (20s)	n	+	++	n	n	n	Mihaljevic and Salopek-Sondi, (2012)
Vaccinum Horibundum	++	n	n	n	++	n	Magnitskiy *et al.* (2011)
Disterigma alaternoides	++	n	n	n	+++	n	

Positive effect of auxins +: low; ++: medium; +++: high; n: not used; 0: no effect

Table 2: Effect of auxins on *Vaccinium Corymbosum*, table drawn up according to the authors referenced. Treatment lasting 20 seconds

Species	Auxin	Concentrations (ppm)				Reference
		1250	2500	5000	10000	
	AIB	+	+	++	++	
Vaccinium *Corymbosum*	EIA	+	+	+	++	Mihaljevic and Salopek- Sondi (2012)
	AIB-Ala	++	++	++	+++	

Positive effect of auxins: +: low; ++: medium; +++: high

In *Vaccinium membranaceum* and *Vmyrtilloides*, the use of ANA in high concentrations did not improve rooting; only with the use of AIB or AIB+ANA was there an increase in the rooting percentage, and concentrations that were too high (10000 and 20000 ppm) may have negatively influenced the final result (Mckechnie *et al.*, 2012) (table 1). In *Vaccinium floribundum* and *Disterigma alaternoides*, both AIB and ANA led to an improvement in rooting, with ANA at the highest concentration (400 ppm) showing the best results (Magnitskiy *et al.*, 2011) (table 1).

In the *Vacciunium corymbosum* species, the auxin AIB-Ala (a conjugate of alanine and indole-butyric acid) was the most effective, and AIB and AIA were also effective in improving the rooting percentage and survival of the cuttings, of the three auxins used, the concentrations of 5000 and 10000 ppm (from 1250, 2500, 5000, 10000 ppm) were the most promising (Mihaljevic and Salopek-Sondi, 2012) (table 2). However, in another experiment, 1000 ppm of AIB (from 500, 1000 and 2000 ppm) showed better results in the same species, and according to a mathematical model from the same experiment, the ideal concentration would be between 641.7 and 712.7 ppm (Celik and Obadas, 2009).

In two experiments with *Vaccinium ashei* the most effective concentrations of

AIB were the maximum used, according to Maragon and Biasi (2013) (table 1) comparing the concentrations of 250, 500, 1000 and 2000 ppm the concentration of 2000 ppm is the one that shows the best results, and according to Vignolo *et al.* (2012) (table 1) who used concentrations of 1500, 3000, 4500, and 6000 ppm, the concentration of 6000 ppm is the one that showed the best rooting.

The varieties used and the place of origin influence the rooting percentage and the survival of the cuttings, for example in *Vaccinum armoreum* the place of origin influenced rooting (Bowerman *et al.*, 2012). In *Vaccinium corymbosum*, the varieties influenced the results, and the auxins favored the varieties that were less suitable for rooting (Mihaljevic and Salopek-Sondi, 2012). In *Vaccinium ashei*, the varieties had an effect on the survival of the cuttings (Vignolo *et al.*, 2012) and their rooting percentage (Maragon and Biasi, 2013).

2.3 Phenology and BBCH

Phenology is the study of events in the animal or plant life cycle, correlated with the surrounding climate. Using a guide to the phenological events or phases of any species can be a fundamental tool for producers to make decisions at the right times regarding cultivation techniques, the use of phytopharmaceuticals and the application of fertilizers (Martinelli *et al.*, 2014).

Over the last 80 years, numerous authors have created different phenological scales for different species. Troitzki (1925) (cited in Meier *et al.*, 2008) studied the correlation between the occurrence and control of the coleopteran *Anthonomus pomorum* L. and the phenological development of apple flower buds. On the basis of this study, Fleckinger (1948) (cited in Meier *et al.*, 2008) described the phenological stages of pome fruits, and this scale was widely used until 1994. Based on the Feekes (1941) scale (cited in Meier *et al.*, 2008), Large (1954) published the first numerical scale for cereals. Zadocks *et al.* (1974) (cited in Meier *et al.*, 2008) presented a readjusted decimal numerical scale for cereals and rice, which is still used by producers of these crops.

As in all branches of science, the various more specific areas of the study of agricultural plants have been working more closely together, and have become more international. The exchange of information between areas presupposes an equal understanding of the terms used. To achieve this equal understanding of terms, there is a need to create an extensive and standardized description of the stages of plant development according to their phenological characteristics and their code (Klingauf cited in Meier, 2001).

The amplified BBCH scale is a system with a uniform decimal code for the proximate phenological phases of all monocotyledonous and dicotyledonous plant species (Hack *et al.*, 1992). The decimal code, which is divided into main and secondary growth stages, is based on the code for cereals developed by Zadocks *et al.* (1974), in order to avoid major changes to this widely used phenological key (Hack *et al.*, 1992). The acronym **BBCH** derives from Biologische Bundesantalt Bundessortenamt and **CHemical** industry.

The development cycle is divided into 10 long and well-defined phases, these phases are described using the numbers 0 to 9 in ascending order from germination to senescence (Hack *et al.*, 1992), (table 3).

Table 3: Main growth phases, table adapted from Hack *et al.* (1992).

State	Description
0	Germination / bud development
1	Leaf development (main branch)
2	lateral branch formation / tillering
3	Stem elongation or rosette growth/development (main branch)
4	Development of vegetative parts for harvesting or vegetative propagation organs
5	Inflorescence emergence (main branch) / ear swelling
6	Flowering (main branch)
7	Fruit development
8	Fruit and seed ripening

Within these divisions there are sub-divisions, called secondary phases, and like the main phases they have the code 0 to 9. These secondary phases are short-lived and evolve within one of the main phases. The combination of the main and secondary phases gives rise to a 2-digit code, the first of which is the main one. Using these two digits makes it possible to precisely define all the phenological phases of most plant species. In some cases, another subdivision is used between the main phase and the secondary phases from 0 to 9, creating a 3-digit code (Hack *et al.*, 1992) (fig.2).

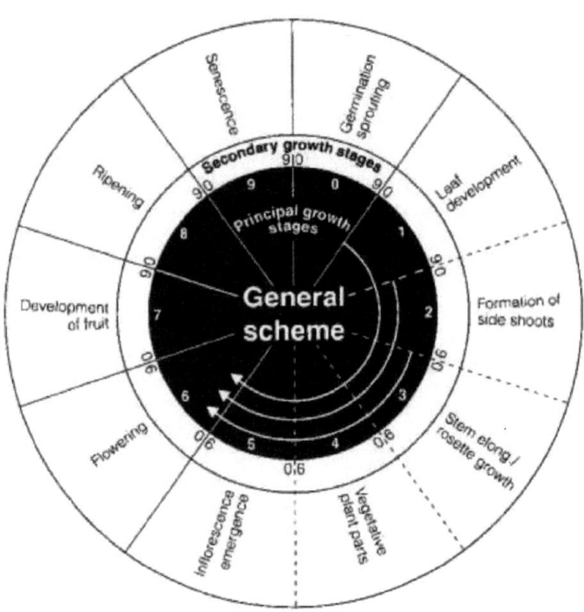

Figure 2: BBCH general scheme, adapted from Hack *et al.*(1992).

This scale has been used to characterize various species of arable and horticultural crops, such as *Zea mays* L., *Vicia faba* L. and *Helianthus annus* L. (Weder and Bieihoider, 1990) and (Lancashire *et al.*, 1991), *Beta vulgaris* L. ssp. *Vulgaris* and *Solanum tuberosum* . (Hack *et al.*, 1993), among many others. It can also be, and has been, used for fruit trees such as *Malus domestica* Borkh., *Pyrus communis* ., *Prunus cerasus* L., *Prunus domestica* L. ssp. *domestica*, *Prunus persica* Batsch., and *Prunus armeniaca* L., (Meier *et al.*, 1994) among others.

2.3.1 Shrimp Phenology, and the Shrimp at BBCH

There are no publications on the phenological states of the shrimp in the BBCH, but there are some studies that characterize the phenology of the species, focusing mainly on the reproductive phase and the influence of investment in the reproductive phases on vegetative growth (Guitian *et al.*, 1997; Zunzunegui *et al.*, 2006; Alvarez-Casino *et al.*, 2010; Oliveira and Dale, 2012).

It is a dioecious plant, but there are also hermaphrodite plants. These different sexual forms have different vegetative and reproductive growth times. Female plants have intense vegetative growth earlier, in pre-flowering and during flowering, but at the time of setting the male plants show greater vegetative growth, while the hermaphrodites have an intermediate behavior (Zunzunegui *et al.*, 2006). There is a low percentage of hermaphrodites, which seems to be explained by the fact that they are less physiologically favored than unisexual plants, although they can be maintained due to natural selection in cases of low population density as they better ensure successful reproduction (Zunzunegui *et al.*, 2006).

According to Guitian *et al* (1997) male plants produce more inflorescences and more flowers than female plants, but the investment in the reproductive phase is greater in female plants, due to fruit production. The fact that male plants produce more inflorescences leads to greater branching and shorter internodes (Alvaréz-Casino *et al.*, 2010).

The growth pattern is similar between male and female plants, but there is a time lag.

The BBCH scale is adapted for short-cycle plants (annuals and biannuals) and long-cycle plants (perennials), such as the shrimp. In these cases, the focus is more on the development of the buds each year, rather than the entire cycle of the plant (from germination to senescence).

As a scale that aims to cover all types of plant species, there are growth phases that don't always apply. For example, several perennial plant species described phenologically on the BBCH scale, such as *Olea europea* (Sanz-Cortés *et al.*, 2001), *Eriobotrya japónica* (Thumb.) Lindl (Martinez-Calvo *et al.*, 1999), *Litchi chinensis* Sonn (Wei *et al.*, 2013), *Annona cherimola* Mill (Cautin and Agusti, 2005), *Diospyros kaki* (Garcia-Carbonell *et al.*, 2002), *Punica granatum* L. (Melgarejo *et al.*, 1996), have in common the fact that some of the main growth phases are not used, such as n^0 "2" - "formation of laterals / tillering" and n^0 "4" - "development of vegetative organs for harvesting or vegetative propagation".

Finn *et al* (2007) proposed an adaptation of the BBCH scale for trees and woody species and in this scale phases 2 (formation of lateral branches) and 4 (development of vegetative organs for harvesting or vegetative propagation) were also omitted, phase 2 being too extensive for the BBCH scale and better described with models of tree architecture, and phase 4 not being relevant in woody species.

As this is a dioecious species, it will be necessary to adapt the scale in order to characterize male and female plants distinctly, and male and female plants can be differentiated in the description of phases "5" and "6" "inflorescence emergence" and "flowering", respectively. The phases relating to the formation, development and ripening of the fruit ("7" and "8") were used exclusively for female plants, as described by Margaret *et al.* (2010) in relation to species of the *Salix* genus.

3. Materials and methods

This study was carried out at the Institute Superior de Agronomia (ISA), University of Lisbon (UL), at the premises of the Instituto Nacional de Investigaçâo Agraria e Veterinaria (INIAV, I.P.) in Oeiras, and in the village of Aldeia do Meco:

1)   Rooting of *C.album* cuttings with two tests: test 1: Importance of genotype, cultivated vs. wild and influence of auxin and test 2: Importance of geographical distribution and type of substrate.

2)   Description of phenological phases and proposal of a BBCH scale for the species *Corema album* (L.) D.Don.

3.1 Shrimp rooting

3.1.1 Rooting site and material

The two rooting trials took place in a glass greenhouse on the premises of the Instituto Nacional de Investigaçâo Agraria e Veterinaria (INIAV, I.P.). The greenhouse has an automatic system for opening the air inlets, which opens when there is a temperature of 28 ^{0}C or more inside. Six raised benches were used in this greenhouse. These benches were equipped with a simultaneous misting irrigation system with drippers that worked every day from 8am to 8pm and watered for 1 minute every 30 minutes to prevent the cuttings from dehydrating. The temperatures and relative humidities that occurred inside the greenhouse between November 3rd and May 31st were between 14.6°C (December 2014) and 25.8°C (May 2015) and 59.1% and 77.2% (fig.3).

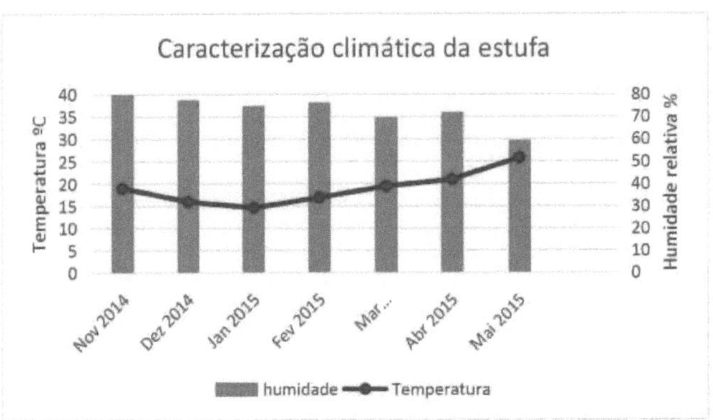

Figure 3: Graph with monthly averages of temperature and relative humidity recorded in the greenhouse, data provided by the INIAV greenhouse complex.

The benches contained hydroponics boxes measuring 89 x 20 x 22 cm and the bottom filled with Leca®10/20 expanded clay (which acts as a drainage element).

Five of the six benches were equipped with an electric heating cable inside the boxes, placed above the Leca®10/20 and a 2 cm layer of substrate, at a depth of around 20 cm. The thermostat of this heating system was programmed to turn off when it reached 24 ^{0}C, and to turn on at 22 ^{0}C.

Two types of substrate were used, the "Fataca" substrate and the Siro®Estaca substrate. The "Fataca" substrate is made up of 3 parts coconut fiber, 2 parts pine bark, 1 part perlite (v/v).

The Siro®Pile substrate consists of Siro®Agro 1 (Humus), selected high-quality peat and Siro®Perlite (appendix 1). The Siro®Pile substrate is referred to as "Siro" throughout the paper.

For trial 1 "Importance of genotype, cultivated vs wild and influence of auxin application" an unheated bench was used, with 16 boxes containing only "Fataca" substrate.

For test 2 "Importance of geographical distribution and substrate type" the remaining 5 benches were used. Two of the benches contained only "Fataca" substrate, another two contained only "Siro" substrate, and finally the last bench was divided so that half of the boxes contained "Fataca" substrate and the other half "Siro" substrate. The various genotypes were divided so that each type of substrate contained the same number of cuttings of each genotype.

3.1.2 Trial 1: Importance of genotype, cultivated vs. wild, and influence of auxin application

The aim was to compare the rooting capacity of wild and cultivated genotypes and their response to different auxin concentrations

There were 4 treatments with 3 concentrations of AIB (indolbutyric acid) and 1 control treatment on cuttings from 2 wild genotypes and 1 cultivated genotype, the cultivated one coming from the same source as 1 of the wild ones.

40 cuttings of each genotype were used for each treatment, giving a total of 160 cuttings per genotype.

The experiment was carried out with all the cuttings placed in the same type of substrate, without heating the root zone and in a protected environment with regular watering.

Dead cuttings were counted 75 days after planting and repeated 105 and 135 days after planting.

The cuttings were removed 165 days after planting, with the number of dead and surviving cuttings recorded. The surviving cuttings were classified on a 0-5 rooting scale, with level 0 having no roots and level 5 having abundant, healthy roots.

3.1.2.1 Areas and vegetation under study

Cuttings were harvested on November 4, 2014, November 21, 2014 and December 11, 2014, from "Aldeia do Meco Cultivada", "Aldeia do Meco Selvagem" and "Quiaios Selvagem", respectively.

Cuttings were taken from one female shrub at each site (referred to as a genotype in this work), for a total of 160 cuttings (with lengths varying between 7.5 and 21.0 cm) per shrub. The choice of genotypes was based on the size of the specimens, in order to have enough size to take the cuttings required to carry out the experiment.

The cultivated genotype came from Aldeia do Meco, but was grown in a substrate and protected environment for 2 years at Herdade da Experimental da Fataca. The wild genotypes came from plants in the respective locations, with the coordinates referenced in table 4.

3.1.2.2 Preparation and treatment of cuttings

The cuttings were not planted on the day of harvest. On the day the cuttings were placed in the substrate, the lower tips were cut off in order to increase auxin absorption. The leaves in the basal zone were removed and in cuttings with more than two branches, the rest were cut off so that there were no more than two branches.

With the help of a ruler, the length of the stake without leaves was measured and the length with leaves was calculated.

The diameter of the stakes was measured with a Mitutoyo ABSolute precision Diagnostic Caliper.

The cuttings were then subjected to 4 types of treatment by soaking them in solutions with different concentrations of AIB: O ppm (control), 5OO ppm, 1OO ppm and 15OO ppm. The solutions were obtained by diluting the auxin in a solution of 25% alcohol (98^0) and 75% deionized water.

To carry out these treatments, 1OO ml of solution of each concentration was poured into a 25O ml beaker. This resulted in a distance of 4.2 cm between the base of the glass and the surface of the solution. This distance represents the treated area of the pile.

The cut ends of the 4O cuttings were dipped in the solution for 2O seconds and this operation was repeated for the 3 auxin concentrations and the control treatment.

3.1.2.3 Pile driving

To place the cuttings in the substrate, the tips of the cuttings previously soaked in the solutions were allowed to dry for a few seconds. The cuttings were arranged in rows on either side of one of the 16 randomly selected bench boxes, marking the rows with the respective genotype and treatment. Each box was divided into three longitudinal sections, one side for the "Aldeia do Meco Cultivado" genotype, the other side for the "Aldeia do Meco Selvagem" genotype, and between them the "Quiaios Selvagem" genotype, totaling 30 cuttings per box.

The cuttings were placed slightly inclined in the substrate to encourage rooting.

The "Aldeia do Meco Cultivado" cuttings were the first to be harvested (November 4, 2014) and were planted the day after the harvest (November 5, 2014). Twenty days later (November 25, 2014) they were transplanted to their final location, and it was on this date that the auxin treatment was carried out.

The "Aldeia do Meco Selvagem" genotype was treated and planted on November 27, 2014 (6 days after harvest), and finally the "Quiaios Selvagem" genotype was treated and planted on December 16, 2014 (5 days after harvest) (table 6).

3.1.2.4 Monitoring, counting and final evaluation

After planting, the watering and heating system was monitored several times. 75 days after planting, the condition of the cuttings was observed, removing any that were dead. The cuttings removed were counted and recorded, and this procedure was repeated 105 and 135 days after planting.

The cuttings were removed 165 days after planting and each one was classified from 0 to 5 according to its root development: 0- no roots, 1- low rooting level, 2- medium low rooting level, 3- medium rooting level, 4- medium high rooting level, 5- high rooting level (fig.4). Dead cuttings were also removed and counted. The length and diameter of all the cuttings removed were measured.

Figure 4: Rooting levels of cuttings, 0- no roots, 1- low rooting level, 2- medium low rooting level, 3- medium rooting level, 4- medium high rooting level, 5- high rooting level.

3.1.3 Test 2: Importance of geographical distribution and substrate type

The aim was to compare the rooting capacity of cuttings from different locations in Portugal, and to compare two types of substrate and their suitability for this stage of plant development.

Cuttings were taken from 10 locations in mainland Portugal and from each location 10 genotypes were chosen at random from among the most vigorous and healthy. From each genotype, 40 to 50 cuttings were selected, thus obtaining between 400 and 500 cuttings from each location.

The cuttings of each genotype were divided into two batches and placed in two types of substrate

The experiment was carried out with heating of the root zone, in a protected environment, with regular watering.

Dead cuttings were counted 75 days after planting and repeated 105 and 135 days after planting.

The cuttings were removed 165 days after planting. When they were removed, the dead and surviving cuttings were noted, and the surviving cuttings were classified on a 0-5 rooting scale, with 0 being no roots and 5 being abundant and healthy roots.

3.1.3.1 Area and vegetation under study

This trial was divided into 6 zones of the country: north, central north, central, central south, southwest and southeast (fig.5). Within each zone there is at least one representative genotype.

Cuttings were collected from the farthest reaches of the country, i.e. the north and southeast. Next, cuttings were taken from the south central zone, the zone with the highest number of genotypes represented. Cuttings from the central and north central zones were harvested on the same day. The south-western zone was the last to be harvested.

In the case of the central zone, at the "Cabo de Carvoeiro" site, only 4 genotypes could be extracted, as only these four had a sufficient number of cuttings with the characteristics required for the experiment.

The coordinates of each location were taken and the date of harvest was recorded (table 4). The number of cuttings and genotypes harvested from each location varied between 34 and 50, with the number of genotypes always remaining the same (10). In the case of Cabo do Carvoeiro, only 4 genotypes were harvested, which is why it was not considered in the statistical treatment (table 5).

Table 4: Location and date of collection of cuttings (in order of harvest). The areas marked with an asterisk were those used in trial 1, and the rest in trial 2.

Zone	Location	Coordinates	Date extracted
South East	Vila Real de Santo Antonio	37° 11'42"N; 7° 24' 36" W	02-Nov-14
South Center *	Aldeia do Meco Cultivated	38° 28' 07"N; 9° 11'09"W	04-nov-14
North	Moledo	41° 51'04"N; 8° 51'56"W	06-Nov-14
South Center	Lagoa de Santo André	38° 07' 11" N; 8° 47' 40"W	13-Nov-14
South Center	Pego	38° 17'31"N; 8° 46' 38"W	13-Nov-14
South Center	Comporta	38° 18' 04" N; 8° 46' 40"W	13-Nov-14
South Center	Meco Village	38° 28' 07"N; 9° 11'09"W	21-Nov-14
South Center *	Wild Meco Village	38° 28' 07"N; 9° 11'09"W	21-Nov-14
North Center	Quiaios	40° 13' 17" N; 8° 52' 22"W	11-Dec-14
North Center*	Wild Okra	41° 13' 17" N; 8° 52' 22"W	11-Dec-14
North Center	Mira	40° 25' 06" N; 8° 47' 07"W	11-Dec-14
Center	Cabo Carvoeiro	39° 28' 07" N; 9° 24' 17"W	11-Dec-14
Southwest	Corporal Sardâo	37° 36' 20"N; 8° 48' 57"W	18-Dec-14

Figure 5: Harvest zones and locations, image obtained using the program: Google Earth Version 7.1.2.2041. Legend: Shaded zones: Green- North; Orange- North center; Yellow- Center; White: South center; Pink- Southwest; Blue- Southeast. Symbols: ? - Harvesting sites for each zone.

Table 5: Total number of cuttings for each genotype of each locality

Location	Genotype									
	1	2	3	4	5	6	7	8	9	10
Vilareal	40	34	40	40	40	40	50	40	50	50
Moledo	44	42	44	48	44	40	42	40	50	50

Location										
Lagoa S.André	48	50	49	48	50	46	46	50	50	50
Pego	44	44	48	46	42	44	50	42	40	40
Comporta	48	44	42	48	44	50	44	44	50	44
Meco Village	50	50	50	50	50	50	50	50	50	50
Quiaios	50	44	48	46	41	47	48	46	50	42
Cabo Carvoeiro	44	50	50	48						
Mira	43	50	42	50	46	42	46	50	42	46
Cabo Sardao	42	40	50	40	46	48	48	46	38	38

3.1.3.2 Preparing the piles

The cuttings were not planted on the day of harvest. On that day, the lower tips were cut off. The leaves in the basal zone were removed, and in cuttings with more than two branches, the rest were cut so that there were no more than two branches.

3.1.3.3 Setting the stakes

The cuttings were divided between the benches so that half the cuttings of each genotype were placed in the "Fataca" substrate and the other half in the "Siro" substrate. The dates on which the cuttings were harvested and placed are described in Table 6, as well as the days that elapsed between these two procedures.

Table 6: Dates of harvesting and planting of cuttings, and days between procedures.

Location	Harvest date	Planting date	Break days
Vila Real de Santo António	02-Nov-14	03-Nov-14	1
Aldeia do Meco Cultivated	04-nov-14	05-nov-14	1
Moledo	06-Nov-14	10-Nov-14	4
Santo Andrè Lagoon	13-Nov-14	14-Nov-14	1
Pego	13-Nov-14	18-Nov-14	5
Comporta	13-Nov-14	19-Nov-14	6

Meco Village	21-Nov-14	25-Nov-14	4
Wild Meco Village	21-Nov-14	27-Nov-14	6
Quiaios	11-Dec-14	16-Dec-14	5
Wild Okra	11-Dec-14	16-Dec-14	5
Cabo Carvoeiro	11-Dec-14	19-Dec-14	8
Mira	18-Dec-14	27-dec-14	9
Corporal Sardâo	18-Dec-14	23-dec-14	5

3.1.3.4 Monitoring, counting and final evaluation

After planting, the watering and heating system was monitored several times. 75 days after planting, the condition of the cuttings was observed and those that were dead were removed and counted. This procedure was repeated 105 and 135 days after planting.

After 165 days after planting, the cuttings were removed and each one was classified from 0 to 5 in terms of rooting: 0- no roots, 1- low rooting level, 2- medium low rooting level, 3- medium rooting level, 4- medium high rooting level, 5- high rooting level (fig.4). The number of dead cuttings was also recorded.

3.1.4 Statistical analysis

For each experiment, a two-factor analysis of variance (treatments) was carried out and Tukey's multiple comparison of means test was performed for $\alpha=0.05$. In the case of rooting percentage, the decimal value of the cuttings survival variable was transformed using the arcsen($\sqrt{x}$) transformation. The chi-square test was used to analyze the level of rooting. The software used for these analyses was Statistix 9 (Analytical Software, Tallahassee, Florida).

3.2 Phenological description of shrimp and introduction to the BBCH scale

The aim was to document and describe the phenological stages of the campania tree and to include this description in the BBCH scale.

Weekly visits were made to a site with various shrimp genotypes, documenting the different stages and evolution observed via photographs and written descriptions.

This description was inserted into the code used in the BBCH scale, following the rules imposed.

3.2.1 Area and plant material under study

The observations of the phenological stages of *C. album* were carried out in the dunes of Praia do Meco in Aldeia do Meco at coordinates: 38° 28' 07" N; 9° 11' 09" W. The dunes are the natural habitat of this species, and are made up of 100% sandy soils. The climate of this location is typically Mediterranean, with a very dry summer and relatively high temperatures, and a very rainy winter with average temperatures of around 11 °C (fig.6).

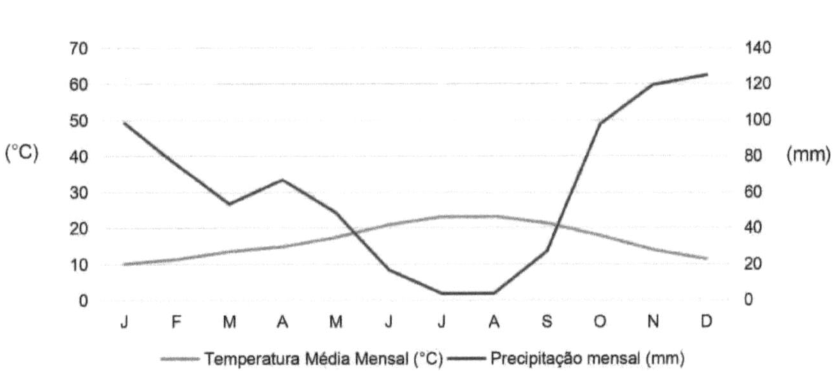

Figure 6: Rainfall graph for Setúbal. Monthly data on average air temperature and average rainfall from a series of climatological normals from 1981 to 2010 (http://www. ipma.pt/en/)

The material under study was two female and two male shrubs of *C.album*, selected because they best represented the different phenological stages within the study site.

3.2.2 Recording and describing the phenological phases and inserting them into the BBCH scale

Weekly visits were made to the dunes of Praia do Meco (Aldeia do Meco) and female and male genotypes were observed, recorded and photographed. The photographs taken focused on different plant organs: the trunk, leaves, branches, buds, inflorescences, flowers and fruit. Images were taken on a weekly basis until the end of the respective flowering periods (January 6th to May 5th), after which the study site was visited every two weeks (May 17th to September 12th). In addition to documenting the evolution of the image, the visits were also used to describe what was observed and the differences compared to previous visits.

The photographs were taken with a Canon EOS 300D camera and an 18-55 mm EFS lens.

Samples were taken from the field to make more detailed observations and photographs at the INIAV, I.P. facilities, with the help of the "Leica DMS1000" magnifying glass and 3D image acquisition *software* "Leica Application Suite V 4.04".

We began observing the phenological stages of the shrimp on January 6, 2015, with the last observation on September 12, 2015.

Initially, a general description was made of what was observed, describing the small details and not just the development of the different phenological phases of the species. In order to summarize and standardize the description of the different phases, it was decided to carry out a description according to the rules of the BBCH scale.

4 Results

4.1 Rooting of camarinha cuttings

4.1.1 Trial 1: Importance of genotype, cultivated vs. wild, and influence of auxin

4.1.1.1 Characterization of the piles

The characterization of the cuttings showed diameters of between 0.7 and 2.5 mm, and average lengths of 15.1, 11.3 and 17.4 cm for the cuttings of the "Aldeia do Meco Cultivado", "Aldeia do Meco Selvagem" and "Quiaios Selvagem" genotypes, respectively (table 7).

Table 7: Characterization of the cuttings: averages, maximums (Max) and minimums (Min) of total length, length of the part with leaves (C/leaf), length of the part without leaves (S/leaf) and diameter of the cuttings.

Origin		Total (cm)	S/leaf (cm)	Leaf length (cm)	Diameter (mm)
Village of	Average	15,1 ± 1,9B	6,9 ±0,9 B	8.2± 1.4 B	1,6 ±0,3 B
Meco	Max	19,0	9,0	12,0	2,3
Cultivated	Min	11,0	5,0	5,0	0,7
Village of	Average	11,3±2,3C	5.6± 1.1 C	5.7± 1.6 C	1,5 ±0,3 C
Meco	Max	19,5	10,0	11,0	2,3
Wild	Min	7,5	4,0	3,0	1,0
	Average	17,4 ± 1,7 A	7,9 ±0,7 A	9.5± 1.3 A	1,8 ±0,3 A
Quiaios	Max	21,0	9,0	12,5	2,5
Wild					
	Min	13,0	6,0	6,5	1,3

Tukey's multiple comparison test of means for α=0.05, different letters in columns indicate statistically different values.

The sizes of the cuttings are significantly different between all the genotypes, with the "Quiaios Selvagem" genotype always being superior to the other two, and "Aldeia do Meco Cultivado" always being superior to "Aldeia do Meco

Selvagem".

4.1.1.2 Survival of cuttings

The dose of auxin applied and the interaction between the dose applied and the genotype had no significant influence on the survival of the cuttings on any of the observation dates. The genotype in turn had a significant influence on all the observation dates (all p=0.0001). The influence of genotype on the survival of cuttings coincides with the work of Mihaljevic and Salopek-Sondi (2012) in which they observed a significant influence of the varieties of *Vaccinium corymbosum* L. (also an Ericacea) on the survival of cuttings.

The "Quiaios Selvagem" genotype with the largest and thickest cuttings showed significantly higher average survival rates on all dates than the "Aldeia do Meco Selvagem" and "Aldeia do Meco Cultivado" genotypes, which in turn showed no significant differences (fig. 7). The fact that this genotype has cuttings of greater length and diameter leads to a greater reserve of carbohydrates and nutrients, which are important factors in successful rooting, since they serve as reserves for the (rootless) cutting until it is able to assimilate what it needs on its own (with the new roots). This fact was highlighted by Costa *et al.* (2013) who compiled a review of various works on vegetative propagation.

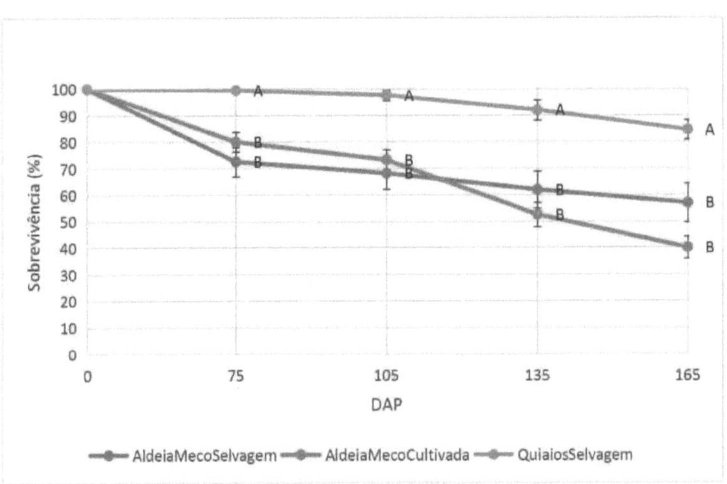

Figure 7: Cumulative survival of cuttings (%), measured 0, 75, 105, 135 and 165 days after planting, of the genotypes "Quiaios Selvagem", "Aldeia do Meco Selvagem" and Aldeia do Meco Cultivada. Tukey's multiple comparison test of means for α=0.05, different letters indicate statistically different values, DAP-Days after planting.

4.1.1.3 Rooting levels

There were only two cuttings with the maximum rooting level (level 5) in the "Aldeia do Meco Selvagem" genotype, which had the smallest cuttings. Of the three genotypes, this was the one that showed the best rooting, with cuttings with all levels of rooting, 54 cuttings (59.3%) with roots, and 31 cuttings (30%) with rooting levels higher than 2 (table 8). The smaller, smaller-diameter cuttings of this genotype are probably cuttings with less differentiated tissues than the larger cuttings, tissues which, according to Costa *et al* (2013), being less differentiated, have a greater rooting capacity. In line with Bowerman *et al.*, (2012), Mckechnie *et al.*, (2012) and Magnitskiy *et al.,* (2011) in the case of Ericaceae, herbaceous cuttings (less differentiated) root better.

The cuttings of the "Quiaios Selvagem" genotype, despite having a high survival rate, had a low rooting capacity, since of the 135 surviving cuttings only 14 (10.4%) had roots, and even these 14 had low rooting levels (12 level 1 and 2 level 2) (table 8). Based on Costa *et al.* (2013), this low rooting rate may be due to the fact that the tissue differentiation of this genotype (with the largest cuttings) is high, leading to a low rooting capacity.

The "Aldeia do Meco Cultivado" genotype also had a low rooting capacity, rooting only 11 cuttings (17.2%). However, it did have some cuttings with high rooting levels (3 with level 3, and 2 with level 4) (table 8).

Comparing cuttings from the same source, "Aldeia do Meco", but under different growing conditions, the results suggest that the conditions to which the plant is subjected influence the rooting capacity of its cuttings, since cuttings from plants subjected to growing conditions do not root as well as those found in the wild.

In the case of the influence of auxin concentration on rooting, only the cuttings of the "Aldeia do Meco Selvagem" genotype depended on the dose of AIB (indolbutyric acid) applied, while for the "Quiaios Selvagem" and "Aldeia do Meco Cultivado" genotypes, the hypothesis of independence cannot be refuted. The cuttings that were subjected to 0 ppm and 1000 ppm showed a higher level of rooting than those that were subjected to 500 ppm and 1500 ppm. These were not the expected results, as in most of the studies on rooting using auxins in species of the Ericacea family, there was an influence of auxin concentration on rooting, either positive (Vignolo *et al.*, 2012, Maragon and Biasi, 2013, Celik and Obadas, 2009, Mihaljevic and Salopek-Sondi, 2012, Magnitskiy *et al.*, 2011) or negative (Mckechnie *et al.*, 2012). Bowerman *et al.* (2012) observed that there was no effect of the different concentrations of AIB applied, a result which coincides with that of this study.

Table 8: Number of cuttings at each rooting level, number of rooted cuttings for each dose of AIB (indolbutyric acid), and p-value for tests of independence between rooting level and dose of hormone applied.

Genotype	No. of surviving stakes	Rooting level[z]						AIB concentration				Probability
		0	1	2	3	4	5	0	500	1000	1500	[y]
Village of Meco Wild	91	37	12	11	24	5	2	18	11	17	8	0,04
Quiaios Wild	135	121	12	2	0	0	0	4	3	4	3	0,84
Village of Meco Cultivated	64	53	6	0	3	2	0	4	2	2	3	0,49
Sum	290	211	30	13	27	7	2	26	16	23	14	

[z]Rooting levels of the cuttings, 0- no roots, 1- low rooting level, 2- medium low rooting level, 3- medium rooting level, 4- medium high rooting level, 5- high rooting level. [y]Probability according to the Chi Square Test.

4.1.2 Test 2: Importance of geographical distribution and substrate type

4.1.2.1 Pile survival

It was observed that in relation to the survival of the cuttings there was a significant influence of the locality from which the cutting came for all the observation dates ($p<0.01$) (table 9), but with an interaction with the substrate for the last date (table 11). These results are in line with the work of Mihaljevic and Salopek-Sondi (2012). Mira, Santo André and Quiaios were the sites with the best results, with survivals of 91.6, 89.2 and 88.1 % respectively, while Moledo had the worst survival with 10.3 % (table 11).

Table 9: Survival averages (%) and respective standard deviations of cuttings from the different locations 75, 105 and 135 days after planting, DAP - Days after planting.

Location		DAP		
		75	105	135

Meco Village	95.0± 9.4 B	90.4± 14.9 B	83.4± 22.9 B
Cabo Sardao	96.1± 6.0 AB	93.5± 9.6 AB	89.6± 15.0 AB
Comporta	98.0± 3.8 AB	97.6 ±3.8 AB	94.5 ±9.5 AB
Mira	99.8± 1.0 A	98.0± 4.2 AB	97,0 ±5,4 A
Moledo	97.5 ±4.8 AB	80.3± 16.0 C	30,0 ±20,2 C
Pego	99.4± 1.6 AB	97.3 ±6.3 AB	89.2± 14.1 AB
Quiaios	98.2± 3.4 AB	96.0± 6.2 AB	92.7± 10.1 AB
Saint André	99.2± 1.7 AB	98.6± 2.0 AB	95.9± 5.0 AB
Vila Real de Santo António	100,0 ±0,0 A	99.6± 1.4 A	96.1 ±4.2 AB
EPM	1,42	2,76	4,13
Location	p = 0,01	p = 0,01	p = 0,01
Location x Substrate	p = 0,67	p = 0,78	p = 0,57

Tukey's multiple comparison of means test for α=0.05, different letters in columns indicate statistically different values, EPM - Standard error of the mean. Interaction not significant.

There was also an influence of substrate type from day 105 (p<0.05) (table 10), and on the last day of observations (day 165) there was an influence of the interaction between locality and substrate (p=0.02) (table 11). This result is in line with that observed by Tchoundjeu *et al.* (2002). These authors observed that there was a 15% difference in the mortality of *Prunus africana* cuttings in different substrates.

Table 10 Survival averages (%) and respective standard deviations of cuttings in the different substrates 75, 105, 135 days after planting, DAP - Days after planting.

Substrate	DAP		
	75	105	135
Fataca	97,7 ±4,7	93.3± 11.6B	82,1 ±26,2 B
Siro	98,5 ±4,6	95,9 ±8,7 A	88,7 ±21,1 A

EPM	0,67	1,30	1,95
Substrate	0,22	p = 0,04	p = 0,01
Location x Substrate	p = 0,67	p = 0,78	p = 0,57

Tukey's multiple comparison of means test for α=0.05, different letters in columns indicate statistically different values, EPM - Standard error of the mean. Interaction not significant.

It was observed that the Siro substrate, when compared to the Fataca substrate, showed significantly higher values for survival of cuttings 105, 135 and 165 days after planting (fig. 8).

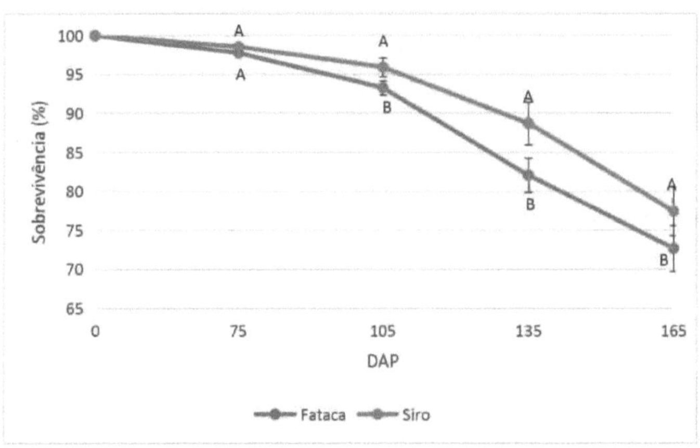

Figure 8: Cumulative survival of cuttings (%), measured 0, 75, 105, 135 and 165 days after planting, in the two substrates used (Fataca and Siro). Tukey's multiple comparison test of means for α=0.05, different letters indicate statistically different values. DAP- Days after planting.

It is worth noting the very low survival of cuttings from the Moledo site in both substrates, which differs greatly from the other sites. The average survival rate for the Moledo site 165 days after planting in the Fataca substrate is 7.9% and in the Siro substrate 12.7% (table 11).

The cuttings with the highest average survival rates on the day they were removed (165 days after planting) were those from Mira on the Siro substrate, those from Quiaios on the Siro substrate and those from Lagoa de Santo André on the Fataca substrate, with averages of 96.5, 93.6 and 93.3 respectively (table 11).

Table 11 Survival averages (%) and respective standard deviations in the interaction between locations and substrate 165 days after planting, DAP - Days after planting.

Location	Substrate	DAP 165
Meco Village	Fataca	66.4± 27.5 B
	Siro	86.4 ±21.4 AB
Corporal Sardâo		75.8± 22.9 AB
	Fataca Siro	85.6± 13.3 AB
		89.4 ±20.5 AB
Comporta	Fataca Siro	83.3± 14.8 AB
Mira	Fataca	86.8± 11.1 AB
	Siro	96,5 ±4,1 A
Moledo		7.9± 12.8 C
	Fataca Siro	12.7± 11.7 C
Pego	Fataca	77.8± 21.9 AB
	Siro	64.9± 17.7 B
Quiaios		82.6± 13.6 AB
	Fataca Siro	93,6 ±9,4 A
Saint André	Fataca	93,3 ±8,1 A
	Siro	85.1± 14.0 AB
Vila Real de Santo António	Fataca Siro	73.5± 15.6 AB
		88.9± 10.8 AB
EPM		7,26
Location		p = 0,01
Substrate		p = 0,04
Location x Substrate		p = 0,02

Tukey's multiple comparison of means test for α=0.05, different letters indicate statistically different values. EPM - Standard error of the mean.

The cuttings with the lowest average survival rates on the day they were removed (165 days after planting) were those from Moledo on both the Fataca and Siro substrates, followed by Pego on the Siro substrate and Aldeia do Meco on the Fataca substrate, with values of 7.9, 12.7, 64.9 and 66.4 %, respectively (table 11).

The percentage drops between the days of observation vary between 0.4 and 8.9 percentage points, except in the case of Moledo, where there were greater drops in the first interval between observations and the second interval between observations, with drops of 17.2 and 50.3, respectively. In the third and final interval between observations (when the interaction begins to have an influence) there were three more dramatic drops in Moledo, Pego and Vila Real de Santo Antonio, with drops of 19.7, 17.8 and 14.9 percentage points respectively.

4.1.2.2 Rooting levels

In general, it was observed that 1626 of the 3096 surviving cuttings had a rooting level of 1 or higher, i.e. around half (52.5%) of the surviving cuttings developed roots. Of the 1626 rooted cuttings, 803 (49.4%) had low and medium low rooting levels, 365 (22.5%) had a medium rooting level, and 458 (28.2%) had medium high to high rooting levels (table 12).

In terms of rooting, the Siro substrate was the one with the best results, accounting for 74.6% of the total rooted cuttings (table 12), even in the case where the values between substrates were closest (Cabo Sardao), with 115 of the 188 cuttings rooted.

To find out whether the level of rooting is independent of the substrate, the Chi-squared test was carried out with a = 0.05 for each location. In all tests, the p-value was less than 0.001, thus rejecting the hypothesis of independence, i.e. the level of rooting is dependent on the substrate in which the cutting was placed. These results are in line with those of Tchoundjeu *et al.* (2002) and Mesèn *et al.* (1996), who compared two substrates and found that they influenced the rooting of cuttings.

Table 12: Number of cuttings in the respective rooting levels, in the different locations and substrates, [z]p-value of the test of independence, [y]p-value of the test of independence between substrate and rooting level.

Location	No. of surviving stakes	Level						Substrate		Probability
		0	1	2	3	4	5	Fataca	Siro	
Meco Village	382	215	62	42	41	18	4	5	162	p=0.000[z]
Cabo Sardao	353	165	62	41	37	31	17	73	115	p=0.000[z]
Comporta	394	184	59	39	52	36	24	47	163	p=0.000[z]
Mira	420	210	72	53	48	25	12	61	149	p=0.000[z]
Moledo	45	21	5	5	3	7	4	1	23	p=0.000[z]
Pego	317	178	49	21	20	30	19	36	103	p=0.000[z]
Quiaios	408	202	50	36	42	45	33	26	180	p=0.000[z]
Lagoon of Santo André	434	170	83	60	70	39	12	90	174	p=0.000[z]
Vila Real de Saint Anthony	343	125	39	25	52	66	36	74	144	p=0.000[z]
Total	3096	1470	481	322	365	297	161	413	1213	
								(p=0.000[y]	p=0.000[y])	

Probability according to Chi square test, [z]Probability of rooting level/substrate according to Chi square test. [y]Probability of rooting level/location

The two substrates differ both in the number of rooted cuttings and in the rooting levels, with rooting in the Siro substrate being higher than in the Fataca substrate. With regard to cuttings with a high rooting level (level 5), the Siro substrate had 150 (12.4%) of the 1213 cuttings rooted, while the Fataca substrate had only 11 (2.7%) of the 413 cuttings rooted (table 13). The low rooting level (level 1) represents 302 (24.9%) of the 1213 rooted cuttings in the

Siro substrate, while this level represents 179 (43.3%) of the 413 rooted cuttings in the Fataca substrate.

To find out whether the level of rooting is independent of the location from which the cuttings came, two Chi-squared tests were carried out with a = 0.05, one for each substrate. In both tests, the p-value was 0.0001, which means that the hypothesis of independence is rejected, i.e. as was seen in the different varieties of *Vaccinium Corymbosum* L. (also an Ericacea) in the work by Mihaljevic and Salopek-Sondi (2012), the level of rooting is dependent on the place of origin/genotype of the cutting.

Based on these results and those of Santos (2013) and Santos *et al.* (2014), it can be concluded that in both types of shrimp propagation there is influence of the origin of the plant material.

The location with the best rooting was Vila Real de Santo António, where 218 (63.6%) of the 343 surviving cuttings were rooted, and the location with the worst rooting was Aldeia do Meco, where 167 (43.7%) of the 382 surviving cuttings were rooted.

In terms of rooting level, the location with the best total rooting (Vila real de Santo António) has the highest percentage of cuttings with levels 4 and 5, with 66 (30.3%) and 36 (16.5%) of the 218 cuttings rooted at the respective levels. The location with the worst total rooting (Aldeia do Meco) is also the location with the lowest percentage of cuttings with a high rooting level, with 18 (10.8%) and 4 (2.4%) of the cuttings rooted at levels 4 and 5, respectively.

Table 13: Number of cuttings showing rooting levels between 1 and 5 in the two substrates under study, and p-value of the test of independence between levels and Locations.

Location	No. of cuttings per rooting level										Total number of rooted cuttings	
	1		2		3		4		5			
	Fataca	Siro	Fataca	Siro	Fataca	Siro	Fataca	Siro	Fataca	Siro	Fataca	Siro

<table>
<tr><td>Meco Village</td><td>4</td><td>58</td><td>0</td><td>42</td><td>0</td><td>41</td><td>1</td><td>17</td><td>0</td><td>4</td><td>5</td><td>162</td></tr>
<tr><td>Cabo Sardao</td><td>30</td><td>32</td><td>19</td><td>22</td><td>14</td><td>23</td><td>8</td><td>23</td><td>2</td><td>15</td><td>73</td><td>115</td></tr>
<tr><td>Comporta</td><td>20</td><td>39</td><td>10</td><td>29</td><td>15</td><td>37</td><td>1</td><td>35</td><td>1</td><td>23</td><td>47</td><td>163</td></tr>
<tr><td>Mira</td><td>23</td><td>49</td><td>19</td><td>34</td><td>10</td><td>38</td><td>1</td><td>20</td><td>4</td><td>8</td><td>57</td><td>149</td></tr>
<tr><td>Moledo</td><td>0</td><td>5</td><td>0</td><td>5</td><td>0</td><td>3</td><td>5</td><td>6</td><td>0</td><td>4</td><td>5</td><td>23</td></tr>
<tr><td>Pego</td><td>17</td><td>32</td><td>11</td><td>10</td><td>4</td><td>16</td><td>3</td><td>27</td><td>1</td><td>18</td><td>36</td><td>103</td></tr>
<tr><td>Quiaios</td><td>15</td><td>35</td><td>8</td><td>28</td><td>3</td><td>39</td><td>0</td><td>45</td><td>0</td><td>33</td><td>26</td><td>180</td></tr>
<tr><td>L.Sto André</td><td>43</td><td>40</td><td>24</td><td>36</td><td>17</td><td>53</td><td>5</td><td>34</td><td>1</td><td>11</td><td>90</td><td>174</td></tr>
<tr><td>V.RSto Antonio</td><td>27</td><td>12</td><td>15</td><td>10</td><td>19</td><td>33</td><td>11</td><td>55</td><td>2</td><td>34</td><td>74</td><td>144</td></tr>
<tr><td>Total</td><td>179</td><td>302</td><td>106</td><td>216</td><td>82</td><td>283</td><td>35</td><td>262</td><td>11</td><td>150</td><td>413</td><td>1213</td></tr>
<tr><td>Level/ Location</td><td></td><td></td><td></td><td></td><td></td><td></td><td></td><td></td><td></td><td></td><td>p=0,00</td><td>p=0,00</td></tr>
</table>

Probability according to the Chi Square Test.

4.2 Shrimp phenology and BBCH scale

4.2.1 Shrimp phenology

The phenological phases observed were described, specifying the time interval in which the different phases were observed. The time frame was summarized in a Gantt chart (fig. 9), in order to make it easier to understand the duration of the different phases.

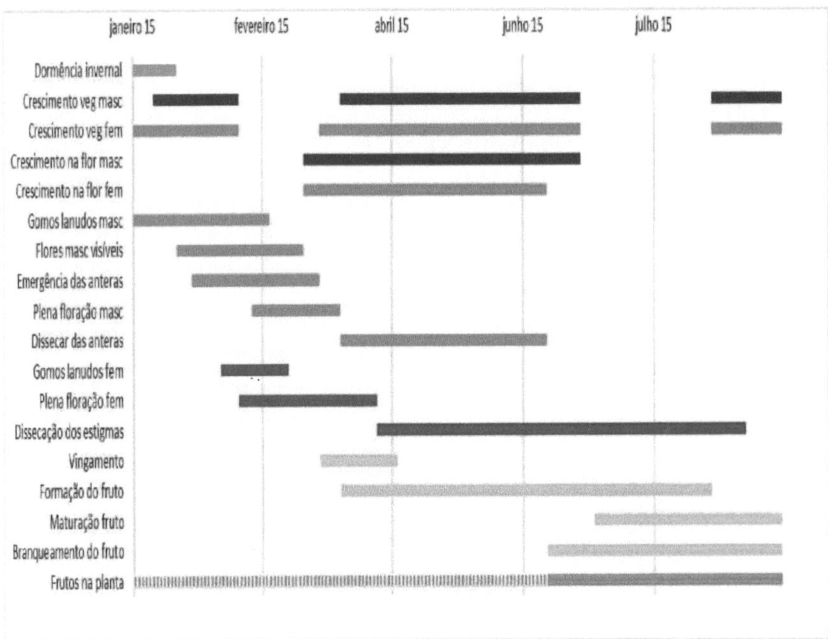

Figure 9: Gantt chart based on observations from January 6 to September 12, 2015. They show winter dormancy in gray, the vegetative phases of male plants in dark green , the vegetative phases of female plants in light green, the male sexual phases in orange, the female sexual phases in red, the yellow phases in yellow.

of the fruit and in blue the existence of fruit on the plant (dashed when it is out of the fruiting season).

No hermaphroditic specimens were observed. One hypothesis for not having observed this phenomenon is that the climate of the site observed (Praia do Meco) is less arid than the sites where the phenomenon has been observed (El Asperillo in Zunzunegui *et al.* (2006) and Vila Real de Santo António in Oliveira and Dale (2012)). Thus, the hermaphroditism of this species seems to be a response to high environmental stress.

Fruits were observed on the plants from January 6th to September 12th (total duration of observations). Although the fruits remain on the plant all year round, they are few in number at the beginning of the year.

Male flowering began in its primordial stages on January 6, 2015, and the anthers began to fall off on March 27, 2015. Female flowering started later, on February 16, 2015, contrary to what was described in Guitan *et al.* (1997), in which the start of flowering was synchronous in both sexes, and ended later, on April 10, 2015. This time difference could lead to the conclusion that the blooms are not synchronous, but the dehiscence of the male plants coincides with the full bloom of the females. This observation coincides with Guitan *et al.* (1997) in which the peak of the two blooms is described as synchronous. The male plants had a greater number of inflorescences per plant than the female plants, and the male plants had more flowers, in line with the results of Guitan *et al.* (1997) and Zunzunegui *et al.* (2006).

The fall of the anthers and stigmas are very long-lasting phases, meaning that both male and female plants retain the traces of their blooms for a long time.

The formation and ripening of the fruit took place between March 27 and September 12, 2015, and seems to have continued at least through September.

Germinated seedlings appeared in the lower zone of male bushes between January 23 and May 29, 2015, and eventually senesced. These seedlings grew without branching and without visible buds (fig.10).

Figure 10: Shrimp seedlings: a) January 23, b) April 04. c) May 29

4.2.2 Leaf phenology

4.2.2.1 Winter dormancy

In the winter dormancy phase, most of the camomina's branches culminate in a pyramidal, black apical bud with thin white stripes (fig.11). When these buds increase in volume, they become more conical and lighter in color, showing a yellowish color with brown stripes on top. This phase was observed from January 6 to 23, 2015 (fig.12).

Figure 11: Bud in winter dormancy

The vegetative buds appear to be the same in both genders.

4.2.2.2 Vegetative growth

Three growth spurts were observed between January 6, 2015 and September 15, 2015 (fig.9).

There are two forms of vegetative growth (fig.13 a and 13b), one (less frequent) in which only the terminal bud, after swelling, develops leaves along the branch, i.e. only one main axis is formed without branches (fig.13b). In the other (more common) form of growth, new axillary buds appear next to the terminal bud, which develop into 3 branches (usually between 2 and 6), in which the original central bud does not seem to develop (fig.13a).

Figure 12: Growth buds

The second growth spurt, called Spring/Summer growth, occurred without buds with visible swelling (fig.13d) in mid-March (March 13, 2015) on the flower structures of the male and female plants. Growth in the vegetative zone of the plants occurred only a week later (March 19, 2015), and in this case there was a difference between the female and male plants, with the female plants starting to grow a week earlier (March 19, 2015) than the male plants, which started growing at the end of March (March 27, 2015). However, there was no substantial difference in the amount of growth between the genders, which does not coincide with Zunzenegui et al. (2006) who stated that the female plant invested less than the male plant in vegetative growth while in the reproductive phase.

This growth takes place in three places in the axillary zone of the central (flower) bud where the leaf primordia appear. Appearing in cone-shaped clusters of numerous leaves in formation, when the tips of the new leaves begin to appear, they are red/dyed, and then turn brown. When there are already visible leaves, the tips are still brown/dyed and the rest of the leaf is light green. As they grow, the leaves separate and elongate, forming a longer, more open cone. You can see from the adult leaves that the young leaves eventually turn a darker green, but this transition is slow.

At this time there is also vegetative growth that occurs in the middle of the inflorescences, both on male and female plants (fig.13c and 13f). This spring/summer growth begins with the leaf primordia, and after a certain amount of growth a pyramidal to conical bud develops, like that of the winter growth but with a beige to orange color and without the brown stripes, it is thought that this bud will eventually stagnate in growth. After this bud has stagnated, new growth occurs in the axils of the bud, or the bud itself resumes growth.

Up until the middle of fruit formation, female plants show greater vegetative growth than male plants, which does not coincide with the work of Zunzenegui *et al.* (2006) in which male plants normally show greater growth from setting. After the middle of fruit formation, vegetative growth seems to be the same between the two genders, coinciding with some years (2000 and 2001) of the study by Zunzenegui *et al.* (2006).

Early summer growth seems to be more long-lasting, although it is not known whether the third and first outbreaks are interconnected due to the limited observation period.

Figure 13: The two types of growth in the different vegetative phases: a) Winter axial growth; b) Winter axis growth; c) Mascuiina inforescence growth; d) Spring/Summer axial growth; e) Spring/Summer axis growth; f) Female

inforescence growth.

The third growth spurt appears to be the type of growth that occurs with visible buds. It began in mid-August (August 16, 2015) and continued until September 12, 2015.

4.2.3 Reproductive phenology

4.2.3.1 Phenology of male flowering

The development of male inflorescences can be divided into 5 phases: flower bud, pre-emergence of the stamens, emergence of the stamens, dehiscence and senescence of the stamens.

Sometimes there is more than one flower bud at each end of the branch. As the inflorescences form, there is vegetative growth in the axils of the flower bud. This growth is small but at the top of the new branches a floral bud is formed which gives a new level of flowering (fig.14).

Figure 14: Two levels of male flowering.

This phenomenon occurs more in male plants and may contribute to the greater number of inflorescences observed in this study and in those of Guitan *et al.* (1997) and Zunzunegui *et al.* (2006), but it also occurs in female plants. Flowering at this second level takes place around one or two weeks after the first, but it is a flowering that goes through the different phases more quickly, extending the total time that the plant is in bloom, increasing the chances of pollination and consequent fruit setting.

Figure 15: Gomo lanudo Masculino

The male reproductive phase lasted 80 days from the cotton bud stage (January 6, 2015) to the end of dehiscence (March 27, 2015) (fig.9).

<u>Fiorai bud phase</u>

The mascuiine floral buds are quite different from the vegetative ones, they are more spherical in shape and larger in volume (fig.15). The development of these buds begins with their swelling and the appearance of fibers that give them the appearance of aigod. The bud continues to swell until the scales open and the next stage begins. The yellow buds were observed on the first day in the field (January 6, 2015) and remained on the plant until the end of February (February 28, 2015), during which time the first and second level aigod buds began to appear.

<u>Stage of pre-emergence and emergence of stamens.</u>

In the pre-emergence phase, convex, cyan-brown and yellowish semi-spheres form under the fibers, giving the appearance of aigod, these fibers disappear from the top of the spheres, and the inforescence looks like a sphere made up of smaller spheres (like raspberries or blackberries). Between 7 and 11 small spheres (fiores) were observed per infiorescence. The spheres begin to fissure, and crimson anthers emerge from inside them (fig.16).

Figure 16: Beginning of stamen emergence

At the emergence stage, the crimson anthers emerge from the cyan-brown tissue of the semi-spheres, and as the stipe unwinds, it pushes the anthers to the outside of the receptacle. The stipe is pinkish white and as it unravels it becomes less pink, turning completely white. The anthers, as they are pushed by the fiiete, begin to darken and turn reddish pink. When the fiiete is almost completely distended, the anthers begin to turn brown and open.

The buds visibly bearing inflorescences were observed in the final half of January (January 23, 2015) and the anthers began to emerge the following week (January 29, 2015). Both these phases lasted 49 days, ending with a gap of one week, with this phase ending in the middle of March (March 13, 2015) and the emergence of the anthers in the final half of March (March 19, 2015). These durations include the first and second levels.

<u>Pollen dehiscence, dissection and fall of stamens</u>

Dehiscence occurred in a short space of time, but was abundant, even creating visible pollen clouds (fig.17). This phase began on February 21, 2015 and lasted until March 27, 2015 with a total duration of 34 days.

Figure 17: Dehiscence

When the stamens elongate, the anthers quickly start to change color, and the brightest colors, such as light pink and dark reddish pink, are short-lived events in the plant. In the dehiscence phase, the filaments turn yellow and the anthers turn brown, splitting to release the pollen, which in the case of the chameleon is carried by the wind (pollination is anemophilous) (fig.17). The dehiscence phase lasts some time, probably because several inflorescences from the same plant develop at different times.

After dehiscence, the stamens dry out more and many end up falling off, detaching themselves from the base of the inflorescence. The fall of the anthers takes a long time (79 days), although at the beginning of this phase (the first 30 to 40 days) there was a large fall of anthers, and in the following days there was a much slower and staggered fall. The fall of anthers coincided with the end of dehiscence, i.e. at the end of March (March 27, 2015).

Once the stamens have dried out, a tuft of "cotton" begins to appear in the center of the inflorescence, and as it disappears, new vegetative growth becomes visible in the center of the inflorescence, with a small cluster of new conical-shaped leaves that then lengthens and widens. Sometimes this new growth only appears when most of the stamens have already fallen off, leaving a cup with a protrusion in the center, which is brown at the base, white in the middle with cotton-like fibers, and at the top you can see the vegetative growth (fig.18).

Figure 18: End of male flowering and start of vegetative growth

4.2.3.2 Phenology of female flowering

Female inflorescences and flowers go through fewer stages than male ones, passing through the shorter woolly bud stage, the stage when the stigmas appear and develop, and the opening of the inflorescence. Female plants also show vegetative growth at the same time as the inflorescence/flowering phases, resulting in the development of two levels of . The vegetative development is much lower than that of male plants.

<u>Woolly bud phase</u>

The woolly bud phase of the female inflorescence (26 days) is much faster than the male phase (53 days), starting in the first days of February (February 9, 2015) and ending in early March (March 7, 2015).

The female flower buds are morphologically identical to the vegetative buds, being of a similar size but more spherical. The buds also go through a cottony phase, although this is less visible than in male plants.

At this stage, the bud increases in volume and fibers form on the outside, giving them the appearance of cotton (fig.19). The flowers pass from the cotton phase directly to the emergence of the stigmas, there is no intermediate phase.

Figure 19: Female woolly bud

<u>Stage of emergence and stigmatic development.</u>

When they appear, the stigmas are dark red with hints of pink. These first stigmas appear in the center of the inflorescence and are short and thick. Then more stigmas start to appear and the first ones start to migrate towards the edges of the inflorescence (the inflorescence starts to open), with several stigmas per inflorescence. The inflorescence is usually made up of 6 to 10 stigmas. As the stigmas develop, they start to become thinner and longer, with a fork at the end.

<u>Inflorescence opening stage</u>

As described in the previous phase, the inflorescence begins to open and concentrates the stigmas at the edges, to the point where there is an empty space (or few stigmas) in the central area. This central space is occupied by a vegetative bud with the conical shape described above.

<u>Vegetative growth on the female inflorescence</u>

A bud appears in the central area of the female inflorescence, giving rise to vegetative growth (fig.20). This is usually after setting. On , some inflorescences have more vegetative growth buds visible (axillary and central zones), but there are others that have no visible vegetative growth, and the vegetative bud may have died or stopped growing. Inflorescences in which the vegetative bud shows more vegetative growth tend to have less setting, or a slower setting and fruit formation process. When there is no vegetative growth, there is a noticeable faster and more vigorous setting.

Figure 20: Central growth of female inflorescence

The flowering phase began a week after the cotton buds appeared, i.e. in the middle of February (February 16, 2015) and ended in the first half of April (April 10, 2015), lasting 53 days. Flowering and fruit formation were observed from February to August, coinciding with the dates indicated by Valdés *et al.* (1987).

4.2.3.3 Reproductive synchrony and hermaphrodite flowers

Female flowering begins and ends later than male flowering, beginning a month later and ending 15 days later, and is faster, lasting 60 days compared to 80 days in male plants. The male dehiscence phase coincides with the female full bloom phase. Although flowering starts tend to be protandric, there is a tendency towards homogamy in the full flowering phase.

During the field observations, no hermaphrodite specimens were observed, contrary to what was observed by Zunzunegui (2006). However, showed male flowers with traces of a female flower - stigmas and ovary - (fig.21a) and female flowers with one anther (fig.21b).

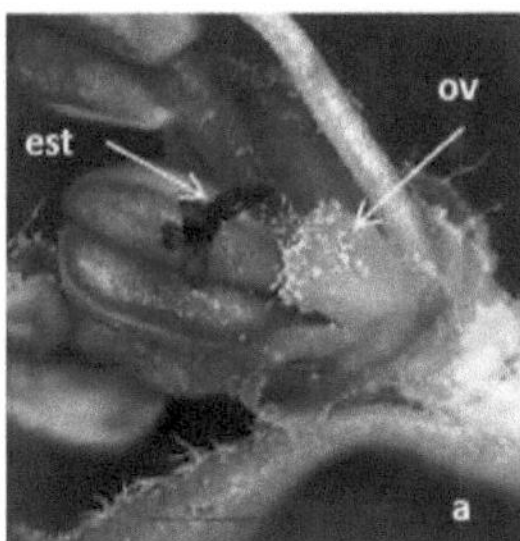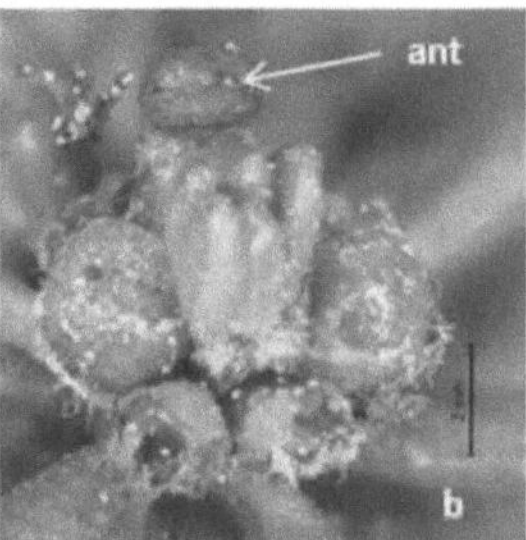

Figure 21: Isolated cases of hermaphrodite flowers. a) male flower with an ovary (ov) and stigma (est); b) female flower with an anther (ant).

4.2.3.4 Revenge

The "fruiting phase" is considered to be from the start of setting until overripe and lasted 163 days. Setting is a short phase lasting 30 days, from the end of March (March 18, 2015) to the end of April (April 19, 2015).

Vining in the camarinha begins with the swelling of the ovary, which has a greenish-yellow color and a concavity at the base of the stigma, which is gradually filled in and the fruit begins to form, with a translucent greenish color. As the fruit develops, it changes to a whiter color, while retaining greenish and yellow tinges. The most common inflorescences bear 3 fruits (fig.22a), but some inflorescences with more than three (up to 6-7) have been seen in the field (fig.22b), while others bear only one or no fruit at all. The setting percentages described by Guitan *et al.* (1997) are between 14 and 28 % at the beginning and between 4 and 17 % at the end, depending on various factors. These authors observed that some fruits after an initial setting did not develop and stagnated in growth, ending up falling or rotting, which may be the cause of the drop in the percentages described.

Figure 22: different types of setting. a) Setting with 3 fruits; b) Setting with 6-7 fruits.

In the setting phase, the normal color of the fruit is green (fig.23a), but some fruits were seen to have a reddish color. This coloration was sometimes visible all over the fruit, but usually only in patches (fig.23b). The change to red was gradual.

There are fruits that remain on the plant in places with two growth spurts above the fruit site, i.e. they have remained on the plant during two growth phases.

Figure 23: Different fruit colors: a: normal color, b: reddish color

4.2.3.5 Fruit development

Newly-formed fruit are similar in shape to ripe fruit, slightly more elongated and more greenish and translucent in color. The color of the newly-formed fruit is a very greenish white, with a little yellow and some transparency. In general, the fruit becomes whiter as it forms and loses its transparency, although there are some fruits that retain a lot of transparency. In terms of shape, it starts out as an elongated sphere, becomes a round sphere and ends up as a flattened sphere. The fruit at the end is a flattened sphere with a dull white color, although there has been some variability in color and shape, with some being more transparent and others more oblong. This variability was also described by Larrinaga (2010) and Guitan (1997).

There are fruits in formation that show a reddish or pinkish color right at the beginning of formation, and as they continue to develop this color darkens and intensifies, and fruits with this color are poorly documented. During the winter, there were more red fruits in relation to the total number of fruits on the plant, so the change to red may be related to the passage of cold periods.

Some of the forming fruits lose their stigma in the early stages of growth, but it is more common for it to remain in the early stages of fruit formation. The fruits remain small for a while, and almost double in size within a week (fig.24).

Figure 24: Fruit growth spurt: a) fruits June 27, 2015, b) fruits July 2, 2015

When the fruit starts to increase in size, there is a noticeable decrease in the vegetative growth of the female plants, compared to the male plants, which at this point only have remnants of flowering and vegetative growth, which is in line with the results of Alvaréz-Casino (2010) and Zunzunegui (2006).

On branches with 3 fruits, which are the most common, it is usual for the fruits on the same branch to increase in size at the same time. On branches with more than 3 fruits, it is not common for all of them to grow at the same time.

The fruit development phase is a very long one (142 days), starting at the end of March (March 27, 2015) and ending in mid-August (August 16, 2015).

4.2.3.6 Fruit ripening

Once the fruit has turned completely white and dull, it starts to develop small brown spots in some areas, which, as well as turning brown, become slightly translucent. The spots are initially small, increasing over time until they eventually cover the entire fruit, turning it light brown and soft. After becoming light brown and translucent, the fruit begins to dull and darken again (fig.25a). After the soft phase, the fruit dries out and shrivels into a sphere with large depressions and the brown becomes more concentrated and darker. Fruits that started out reddish, or turned red as they formed, darken in color as they ripen and turn red in the soft stage (fig.25b), and when they dry out, they blacken and become a black sphere with depressions that have reddish edges.

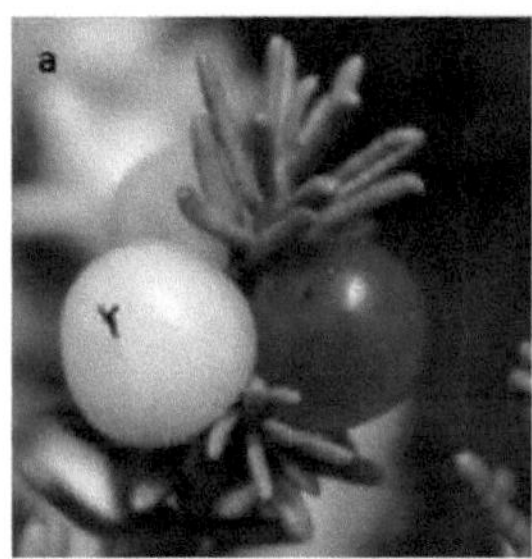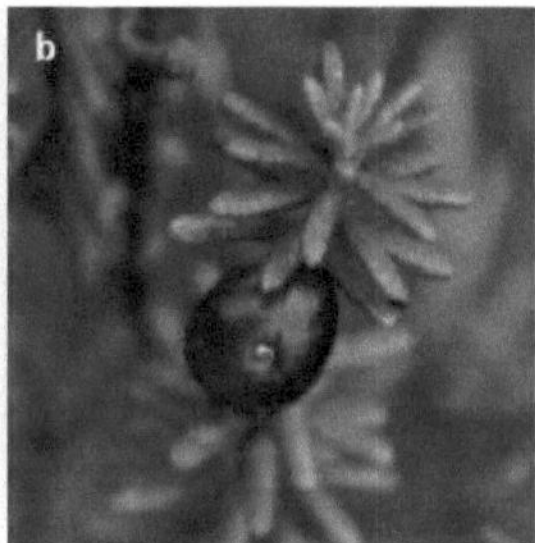

Figure 25: Overripe fruit of the two colors: a) normal, brownish color, b) reddish color.

The fruit was observed on the plant on all the observation dates, from January 6th (2014 fruit) to September 12th (2015 fruit). However, their presence became less abundant in the months of November/December, which is in line with Valdés *et al.* (1987) who stated that the fruit was only observed on the plant until the beginning of winter.

4.2.4 BBCH scale proposal for the *Corema album* species

As there is no BBCH scale for *Corema album*, 7 phenological phases are proposed, and within these 3 phases with sub-phases (table 14). The proposal by Finn *et al* (2007) was adopted, so phases 2 and 4 were not included, as this is a perennial plant. Stage 1 (leaf development) was divided into two sub-phases due to the differences observed in growth at different times of the year (one in winter and the other in spring/summer). In the case of phases 5 and 6, subphases were created due to the fact that this is a dioecious plant, i.e. a subphase was created for male plants and another for female plants, preferring this solution to that of Margaret *et al.* (2010) in which the different phases of inflorescence and flowering of male and female plants of the genus *Salix* spp. were described together and with the same code.

Table 14: Main growth phases of *Corema album*

Phase	Description
0	Bud development
1.1	Leaf development (main branch) Winter
1.2	Leaf development (main branch) Spring
3	Elongation of the vegetative branch / shoot development
5.1	Emergence of the male inflorescence
5.2	Emergence of the female inflorescence
6.1	Male flowering
6.2	Female flowering
7	Fruit development
8	Fruit and seed ripening

Phase 0: bud development

00 Winter dormancy (fig.26).

The bud has a grayish-black color and a rather pointed conical shape.

01 The bud begins to swell (fig.26).

The bud has a pyramidal to conical shape and a black coloration, and begins to separate the black filaments, showing a beige/yellow color inside.

02 Leaf buds begin to thicken and open (fig.26).

The bud opens more and begins to show a white/beige coloration in the center.

03 End of bud swelling (fig.26).

At the end of the bud's swelling, it has a brown color with white/beige in the interstices.

07 Start of bud break: green tips of the leaves start to become visible (fig.26).

The green tips of the leaves begin to emerge from the center and are light green, still covered in filaments that give them a cottony appearance.

09 Bud opening, leaf primordia clearly visible (fig.26).

The tips of the leaf primordia are very small (less than 1 mm), and for this reason a magnifying glass was used for this stage and the one described above (07 and 09).

Figure 26: **01** - Beginning of bud swelling; **02** - Leaf buds begin to thicken and open; **03** - End of bud swelling; **07** - Beginning **of** bud opening; **09** - Bud opening.

Phase 1: Leaf development

During the field observations, there was a difference between the winter and spring/summer growths in terms of budbreak and leaf development.

Because of this, it was divided into two sub-phases: phase 1.1: Winter leaf development and phase 1.2: Spring and summer leaf development.

The vegetative buds in *Corema album* separate the leaves into groups of three, growing attached to the bud and with the tips welded together, these tips then separate and move away from the bud, opening.

Phase 1.1: Winter leaf development

At this stage o leaf development is considered to start from a clearly visible vegetative bud, the leaves form on the buds and separate and develop below it.

110 First leaf tips are visible, leaves are beginning to emerge (fig.27).

The leaves are light green and are still welded together at the end and in the center of the bud.

111 First leaves visible: first leaves separate from the branch (fig.27).

The leaves separate at the top and move away from the bud

112 Formation of the second set of light green leaves (fig.27).

The new sets of leaves appear below the previous set, with each leaf of this new set appearing between two of the previous set, meaning that when you look at them from above they look like sets of 6 leaves.

113 Formation of the third set of leaves (fig.27).

The formed branch begins to show a reddish color and the presence of white filaments, which gives it a cottony appearance.

114 Formation of the fourth set of leaves (fig.27).

The leaves begin to develop a yellow petiole and the central vein of the leaf becomes visible.

115 Formation of the fifth set of leaves (fig.27).

The central vein of the leaf becomes more visible.

116 Formation of the sixth set of leaves (fig.27).

The central vein of the leaf is formed.

117 Formation of the seventh set of leaves (fig.27).

The branch begins to show a duller color, with the reds turning brown and the yellows turning white/grey.

118 Formation of the eighth set of leaves (fig.27).

119 Nine clusters of light green leaves are visible (fig.27).

Figure 27: The different sub-phases of the winter leaf development phase; **110** - First leaf tips are visible; **111** - First leaves separate from the branch; **112** - Formation of the second set of leaves; **113** - Formation of the third set of leaves; **114** - Formation of the fourth set of leaves; **115** - Formation of the fifth set of leaves.**116** - Formation of the sixth set of leaves; **117** - Formation of the seventh set of leaves; **118** - Formation of the eighth set of leaves; **119** - Nine

sets of leaves formed.

Phase 1.2: Early and summer leaf development

During the hottest periods, vegetative growth occurs rapidly, with the newly formed bud giving rise to clusters of leaves in the axillary zone of the flower buds. These clusters of leaves appear in the shape of inverted cones and continue in this shape until the leaves separate, presenting the appearance of a normal branch.

120 First leaf apexes are visible, leaves are beginning to emerge (fig.28).

The leaves are brownish red at the tips and light green on the rest of the leaf.

121 First set of visible leaves (fig.28).

The first leaves begin to elongate away from the center, with the tips of the next leaves appearing in the middle of the first set of leaves. The tips are brownish red and the rest are light green.

122 Second set of leaves visible (fig.28).

The second set of leaves appears in the middle of the previous set, with the new leaves remaining attached to each other and the previous leaves more separated from the center.

123 - Third set of leaves visible (fig.28).

Leaves start to lose their red color.

124-Fourth set of visible leaves (fig.28).

Leaves with almost no red color only at the end of the tip, and the separation of the middle of the leaf starting to form.

125-Fifth set of visible leaves (fig.28).

Central vein of the leaf practically formed with a yellowish color inside.

126-Six sets of leaves visible (fig.28).

Central vein of the leaf formed with the color changing to /beige inside.

127-Seventh set of leaves visible (fig.28).

The central vein of the leaf is white.

128-Eighth set of visible leaves (fig.28).

129-Nine sets of visible leaves (fig.28).

Figure 28: **120-** Leaves starting to emerge, **121 -** First set of leaves visible, **122**

- Second set of leaves visible, **123** - Third set of leaves visible, **124** - Fourth set of leaves visible, **125** - Fifth set of leaves visible, **126** - Sixth set of leaves visible, **127** - Seventh set of leaves visible, **128** - Eighth set of leaves visible, **129** - Nine sets of leaves visible.

Stage 3: Branch development

31 Beginning of branch growth: the axis of the developing branches begins to become visible and they are 10% of their final length (fig.29).

Branches have a reddish-yellow color.

33 Branches are 30% of their final length (fig.29).

Branches have a reddish-yellow color

35 Branches are 50% of their final length (fig.29).

Branches have a brownish-yellow color.

37 Branches are 70% of their final length (fig.29).

Branches have a brownish-yellow color

39 Branches are 90% of their final length (fig.29).

Branches have a brownish-grey color.

Fig. 29: Phenological stages of phase 3 branch development: **31** - Start of branch growth; **33** - Branches are 30% of their final length; **35** - Branches are 50% of their final length; **37** - Branches are 70% of their final length; **39** - Branches are 90% of their final length.

Stage 5.1: Emergence of the male inflorescence

510 Closed floral bud covered with white filaments (fig.30).

The bud looks similar to the vegetative bud, although it is larger and more spherical, with a yellowish color covered in white filaments that give it a cottony appearance.

511 The bud begins to swell and the scales begin to separate (fig.30).

The white filaments begin to disappear and the bud starts to split.

512 Bud begins to lose its cottony appearance and individual flower buds with brown scales form (fig.30).

The cracked areas begin to develop individually, creating small, somewhat pyramidal protrusions, and at this stage they are still covered with cotton.

513 Buds continue to lose their cottony appearance and the individual flowers begin to separate more from each other, being covered in brown scales and some cotton (fig.30).

The protrusions begin to take on a more spherical shape and continue to lose cotton, with the area below (which has a brownish-yellow color) the filaments being more visible than the filaments themselves.

514 Buds already have less cotton than scales and individual flowers are almost completely visible (fig.30).

Individualized scales, but not yet completely open, with a light brown color and a cottony appearance.

515 Male flowers are clearly visible and individualized with dark brown scales (fig.30).

The small spheres formed are clearly visible inside a large sphere (the bud), and are a slightly darker brown than in the previous phase, but still light brown.

Figure 30: States of phase 5.1 Emergence of the male inflorescence: **510 -** Closed flower bud; **511 -** Beginning of bud swelling; **512 -** Bud begins to lose

its cottony appearance; **513** - Bud continues to lose its cottony appearance; **514** - Bud already has less cotton area; **515** - Male flowers are clearly visible and individualized.

Stage 5.2: Emergence of the female inflorescence.

520 Closed floral bud covered with white cotton (fig.31).

The female flower bud is more similar to the vegetative one than the male flower bud, as it also has a conical, slightly rounded shape.

525 Flower bud retains the appearance of cotton and individual stigmas begin to emerge in the center of the bud (fig.31).

The bud increases slightly in volume and soon the ends of the first stigmas appear in the central area of the bud.

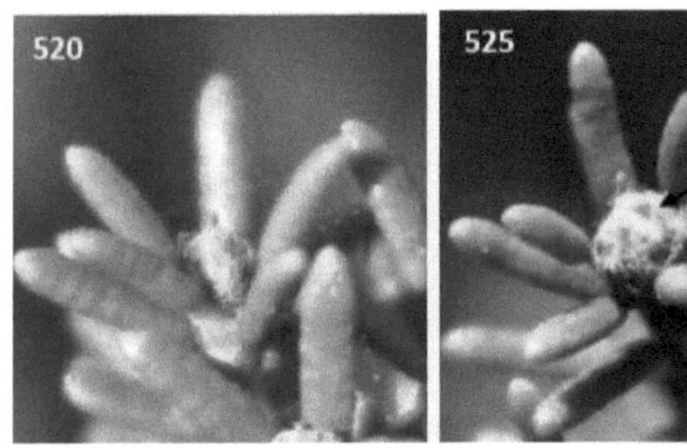

Figure 31: States of phase 5.2 Emergence of the female inflorescence:

520 - Floral bud closed and covered with white cotton; **525** - Floral bud retains cotton appearance and individual stigmas begin to emerge in the center of the bud; **Est** - Stigma emerging.

Stage 6.1: Male flowering

610 The scales open and the first red anthers begin to emerge (fig.32).

The light brown scales begin to split and open, and pinkish red anthers appear in the interstices.

611 Start of flowering: 10% of anthers emerged (fig.32).

The scales continue to split and at least 10% of them already have visible anthers and the red begins to fade and the pink becomes more apparent.

613 Initial flowering: all the anthers are visible and the fillet is beginning to elongate (fig.32).

All scales open, anthers visible, beginning of fillet elongation in some anthers, which retain their reddish-pink color.

615 Anthers begin to dehiscence (fig.32).

When the anthers are about to dehiscence, they show a yellowish color in the splitting zone. When the fillet is at the beginning of the elongation phase, it has a pale pink color.

617 Beginning of the release of the pylon (fig.32).

At the beginning of pollen release, the fillet is white and the anthers are yellow-brown to light brown.

619 End of flowering anthers turn brown (fig.32).

The anthers turn dark brown and the filaments turn a dull yellow and begin to senesce and fall off.

Figure 32: States of phase 6.1 Male flowering: **610** - Scales open; **611** - Flowering begins; **613** - Initial flowering; **615** - Anthers begin to dehiscence; **617** - Pollen release begins; **619** - End of flowering anthers turn brown.

Stage 6.2 Female flowering

620 Beginning of the emergence of the stigmata (fig.33).

The woolly-looking flower bud begins to be pierced by the bright red stigmas, which appear in the center of it.

621 Start of flowering: stigmas emerging (fig.33).

The stigmas emerged in sets of 3 per ovary, with a bright red, somewhat dark color.

623 Initial flowering: central area with several stigmas that begin to move away from the center (fig.33).

The stigmas continue to appear in the center of the bud and push the previous ones away from the center, and the first stigmas begin to elongate.

625 Stigmas distributed throughout the bud, with the area in the center starting to run out of stigmas (fig.33).

At this stage, the stigmas are almost at the periphery and continue to elongate.

627 Elongated stigmas, beginning of the appearance of a vegetative meristem in the center of the inflorescence (fig.33).

The stigmas have already reached their final size, becoming thinner and with a fork at the apex, showing a dark red color. Vegetative growth begins to appear in the center.

629 End of flowering: the beginning of setting (fig.33).

In the basal zone of the stigmas, they begin to set.

Figure 33: Stages of phase 6.2 Female flowering **620** - Stigmas begin to appear; **621** - Flowering begins: stigmas emerge; **623** - Initial flowering; **625** - Stigmas distributed throughout the bud; **627** - Stigmas elongate; **629** - Flowering ends

Stage 7 Fruit development.

70 Avenging (fig.34).

Fruit formation begins with the swelling of the ovary, which becomes spherical and has a central cavity near the insertion of the stigmas, with a light green color.

71 Fruits at 10% of final size (fig.34).

The fruit increases in volume, the central cavity disappears and the color starts to become duller and whiter.

72 Fruits at 20% of final size (fig.34).

The fruit increases in volume and turns a greenish-white color.

73 Fruit at 30% of final size (fig.34).

The fruit continues to increase in volume and turns grayish white.

74 Fruit at 40% of final size (fig.34).

The fruit continues to increase in volume and the grayish white begins to lighten, the growth more pronounced on the longitudinal axis, becoming a more oblong fruit at this stage.

75 Fruits at 50% of final size (fig.34).

The fruit continues to increase in volume and the grayish white lightens, the fruit remains oblong at this stage.

76 Fruits at 60% of final size (fig.34).

The fruit continues to increase in volume and the color turns white, and the fruit begins to grow sideways, becoming spherical at this stage.

77 Fruit 70% of final size (fig.34).

The white fruit continues to increase in volume and continues to grow sharply along the transverse axis, being practically spherical at this stage.

78 Fruit 80% of final size (fig.34).

The white fruit continues to increase in volume and continues to grow sharply along the transverse axis, beginning to take the shape of a flattened sphere.

79 Fruit 90% to 100% of final size (fig.34).

The white fruit reaches its final size, taking the definitive shape of a sphere slightly flattened along the longitudinal axis.

Figura 34: States of phase 7 Fruit development: **70 -** Setting; **71 -** Fruits at 10% of final size; **72 -** Fruits at 20% of final size; **73 -** Fruits at 30% of final size; **74** - Fruits at 40% of final size; **75 -** Fruits at 50% of final size; **76** - Fruits at 60% of final size; **77 -** Fruits at 70% of final size; **78 -** Fruits at 80% of final size; **79** - Fruits at 90% to 100% of final size.

Stage 8 Fruit and seed ripening 81 Fruit ripening (fig.35).

The white fruit reaches its final size, taking the definitive shape of a sphere slightly flattened along the longitudinal axis.

85 Advanced maturation (fig.35).

The fruit begins to show small areas of yellow and light brown on the skin.

87 Beginning of senescence (fig.35).

The fruits begin to soften and create small striations

89 Senescence of the fruit (fig.35).

The whole fruit is a slightly darker brown color.

Figura 35:   Stage 8 Fruit and seed ripening: **81** - Beginning of fruit ripening; **85** - Advanced ripening; **87** - Fruits begin to soften; **89** - Complete ripening

5. Conclusion

From the rooting tests carried out, it can be concluded that the camomina genotype influenced the rooting capacity of the cuttings, while the auxin concentrations used had no significant influence.

With regard to the substrates, it was found that the two substrates had different results in terms of both the number of rooted cuttings and the rooting levels. Rooting in the Siro substrate was higher than in the Fataca substrate, which is therefore the most suitable of the two for propagating this species. The origin of the cuttings also influenced their rooting capacity

The BBCH scale summarizes well the main phases of the vegetative and reproductive development of the shrimp. However, this species has many details that are difficult to match with a generic scale.

<u>In more detail, with regard to the propagation aspect, we can</u> conclude:

<u>Trial 1</u>: As for the "type of genotype", the results from Aldeia do Meco indicate an influence of the genotype on rooting, as the wild genotype obtained better results than the cultivated one. This suggests that the type of conditions to which the plant is subjected influences its rooting capacity. The Wild Quiaios genotype was the one that showed the worst result, so the genotype seems to have a greater influence than the cultivation conditions.

The largest lengths and diameters of the cuttings coincide with the best survival and the worst rooting, which leads us to the conclusion that the larger cuttings had more reserves, but had more lenhified tissues o which reduced their rooting capacity.

There was no significant effect of the different auxin concentrations in answering question 3. This could be due to a lower intrinsic rooting capacity in the cuttings, since the original inoculants did not show the best results in trial 2. Another hypothesis is that in test 1 the benches were not heated, leading to less rooting than in test 2.

Trial 2: The results obtained in this trial make it possible to answer question 1) "Do the origin and physiological characteristics influence the survival and/or rooting of cuttings?", concluding that the origin of the cuttings influences the propagation capacity of this species. These results could be due to differences in climate and natural selection in the different locations, which have conditioned the development of the branches (in the case of propagation by cuttings), fruits and seeds (in the case of propagation by seeds), which could lead to a different response capacity.

The results of this trial provided a positive answer to question 2) "Do different substrates influence the survival and/or rooting of cuttings?": One hypothesis to explain the differences between substrates is their composition. The Siro substrate proved to be more suitable for this method of propagation. Of the two substrates, the Fataca substrate is more similar in composition to the shrimp's habitat; perlite simulates sand in terms of water drainage and air supply to the roots, pine bark is a characteristic component of the shrimp's habitat, and coconut fiber has some salinity (simulating coastal areas), which would lead us to predict better results with this substrate.

The more positive results with the Siro substrate may be due to the fact that it contains substances that promote rooting (appendix 1), as it is a commercial substrate for cuttings, the composition of which is not fully described.

Although the Fataca substrate simulates the characteristics of the shrimp's habitat well, it may not be very conducive to rooting.

<u>In terms of phenology and inclusion in the BBCH scale</u>

Male and female flowering are synchronous. However, vegetative growth is asynchronous, which is due to the different timing and need for investment of the plant's resources in the formation of the different reproductive organs, answering questions 4 and 5.

This species has invested heavily in reproduction, both in terms of the number of fruits produced and the time devoted to reproductive growth, a fact that is closely linked to the harsh conditions of its habitat. It is these conditions that promote strong vegetative growth at the same time as flowering, increasing its duration and the number of flowers, in order to increase the likelihood of reproduction.

In relation to question 6) "Does the BBCH scale serve to describe all phenological events?" This work shows that this scale summarizes the main phases and aspects very well, however, some more specific details are not covered. Given that the observations were only carried out in Aldeia do Meco, some of the details and particularities described in the work may be of greater importance in other locations, which could lead to an adaptation of the scale proposed here.

<u>Interaction of the two strands</u>

In order to demonstrate the usefulness of the two strands of work analyzed, the question 7) "At which phenological stages should we take cuttings with the highest probability of survival/rooting?" was posed, and it was decided that these would be the stages when the seventh set of leaves and the central vein of the leaves are already formed (stages 117-119 and 127-129) and when the branch is between 50% and 70% of its final size (stages 35-37), because it is at these stages that we see cuttings with the morphological characteristics that showed the best results in this work (sufficient size, but not too lignified), but not too lignified.

Future work

Vegetative propagation

Given that the factor "physiological state of the cutting to be rooted" was the one that most influenced rooting success, experiments should be carried out with an extensive characterization of the cuttings to be rooted, separating them into groups with similar characteristics in terms of length, diameter, percentage of carbohydrates and nutrients, relating them to rooting rates.

New trials with different concentrations and types of auxin are needed, evaluating their interaction with the different locations and root heating. The Vila Real de Santo António location which showed the highest rooting rate (maximum) and the Moledo location with the lowest rooting rate (minimum) should be used.

Phenology

As no hermaphrodite plants have been observed in Aldeia do Meco, but they have been described in the bibliography, it would be of great interest to extend the phenological study to locations where hermaphrodite plants have been observed and include them in the BBCH scale proposed here.

As this work was only carried out in one location and in one year, it will be important to continue the phenological characterization of the shrimp, both in the same location in different years and in other locations.

The use of the BBCH scale proposed in this work will help to delineate the optimum times for cultural operations, once the species has been introduced into culture.

The endemic *Corema album* (never commercially exploited) is a very promising species of small fruit, given its unique characteristics of color, taste and benefits for human health.

6. References

Acosta, M., Oliveros-Valenzuela, M. R., Nicolas, C., and Sanchez-Bravo, J. (2009) - Rooting of carnation cuttings. The auxin signal. Plant Signaling & Behavior 4 (3): 234-236.

Ahkami, A. H., Lischewski, S., Haensch, K. T., Porfirova, S., Hofmann, J., Roletschek, H., Melzer, M., Franken, P., Hause, B., Druege, U., and Hajirezaei, M. R. (2009) - Molecular physiology of adventitious root formation in Petunia hybrid cuttings: involvement of wound response and primary metabolism. New Phytologist 181: 613-625.

Altamura, M. M. (1996) - Root histogenesis in herbaceous and woody explants cultured in vitro. A critical review. Agronomie 16: 589-602.

Alvaréz-Casino, L., Zunzunegui, M., Diaz Barradas, M.C., and Esquivias, M.P. (2010) - Gender-specific costs of reproduction on vegetative growth and physiological performance in the dioecious shrub *Corema album.* Annals of Botany 106 (6): 989-998.

Arnaud, C., Bonnot, C., Desnos,T., and Nussaume, L. (2010) - The root cap at the forefront. Comptes Rendus Biologies 333 (4): 335-343.

Bellini, C., Pacurar, D. I., and Perrone, I. (2014) - Adventitious Roots and Lateral Roots: Similarities and Differences. Annual Review of Plant Biology 65: 639-666.

Bowerman, J.R., Spiers, J.D., Coneva, E., Tilt, K.M., Blythe, E.K., and Marshal, D.A. (2012) - Propagation of Sparkleberry (Vaccinium arboreum) Improved Via Cutting Type. Acta Hort. (ISHS) 1014: 385-388.

Calvino-Cancela, M (2004) - Ingestion and dispersal: Direct and indirect effects of frugivores on seed viability and germination of Corema album (Empetraceae). Acta Oecologica. 26: 55-64.

Cautin R., and Agusti, M. (2005) - Phenological growth stages of the cherimoya tree (*Annona Cherimola* Mill.). Scientia Horticulturae 105: 491-497.

Costa, C. T., Almeida, M. R., Ruedell, C. M., Schwambach, J., Maraschin, F. S., and Fett-Neto, A. G. (2013) - When stress and development go hand in hand: main hormonal controls of adventitious rooting in cuttings. Frontiers in Plant Science 4: Article 133- doi: 10.3389/fpls.2013.00133

Celik, H., and Obadas, M.S. (2009) - Mathematical modeling of indole- 3butyric acid aplications on rooting of nothern highbush blueberry (Vaccinium corymbosum L.) softwood-cuttings. Acta Physiol Plant 31: 295-299.

Costa, C.A.(2011) - Factors conditioning the dispersal and recruitment of shrimp in dune systems. [Online] Lisbon: Faculty of Sciences. 42p. University of Lisbon. Dissertation for a Master's Degree in Conservation Biology [Consult. 21 Jan. 2015] Available at http://repositorio.ul.pt/bitstream/10451/3055/1/ULFC090655_tm_C atarina_Costa.pdf

De Klerk, G. J., Guan, H., Huisman, P., and Marinova, S. (2011) - Effects of phenolic compounds on adventitious root formation and oxidative decarboxylation of applied indoleacetic acid in Mallus "Jork 9". Plant Growth Regulation 63 (2): 175-185.

Dello Ioio, R., Linhares, F. S., Scachi, E., Casamitjana-Martinez, E., Heidstra, R, Costantino, P., and Sabatini, S. (2007) - Cytokinins determine Arabidopsis root-meristem size by controlling cell differentiation. Current Biology 17 (8): 678-682.

Dello Ioio, R., Nakamura, K., Moubayidin, L., Perilli, S., Taniguchi, M., Morita, M. T., Aoyama, T., Constantino, P., and Sabatini, S. (2008) - A genetic framework for the control of cell division and differentiation in root meristem. Science 322 (5906): 1380-1384.

Dewitte, W., and Murray, J. A. (2003) - The plant cell cycle. Annual Review of Plant Biology 54: 235-264.

Doener, P. (1998) - Root development: Quiescent center not so mute after all.

Current Biology 8 (2): R42-R44.

Druege, U., Zerche, S., and Kadner, R. (2004) - Nitrogen- and storage- affected carbohydrate partitioning in high-light-adapted Pelargonium cuttings in relation to survival and adventitious root formation under low light. Annals of Botany 94 (6): 831-842.

Druege, U., Zerche, S., Kadner, R., and Ernst, M. (2000) - Relationship between nitrogen status, carbohydrate distribution and subsequent rooting of Chrysantheum cuttings as affected by pre-harvest nitrogen supply and cold-storage. Annals of Botany 85 (5): 687-701.

Finn, G.A., Strazewski, A.E., and Peterson, V. (2007) - A general growth stage key for describing trees and woody plants. Annals of Applied Biology 151 (1): 127-131.

Garcia-Carbonell, S., Yague, B., Bleiholder, H., Hack, H., Meier, U., and Augusti, M. (2002) - Phenological growth stages of the persimmon tree (*Dyospyros kaki*). Annals of Applied Biology 141: 73- 76.

Garrido, G., Guerrero, J. R., Cano, E. A., Acosta, M., and Sanchez-

Bravo, J. (2002) - Origin and basipetal transport of the IAA responsible for rooting of carnation cuttings. Physiologia Plantarum 114: 303-312.

Gonzalez, G. A. L (2001) - Los Arboles y Arbustos de la Peninsula Ibèrica e Islas Baleares, Mundi-Prensa, Madrid 2: 1727.

Guerrero, J. R., Garrido, G., Acosta M., and Sanchez-Bravo (1999) - Influence of 2,3,5-triiodobenzoic acid and 1-N-naphthylphthalamic acid transport in carnation cuttings: relationship with rooting. Plant Growth Regulation 18: 183-190.

Guitian, P., Medrano, M., and Rodriguez, M. (1997) - Reproductive biology of *Corema album* (L.) D. Don (*Empetraceae*) in the northwest Iberian Peninsula. Acta Botanica Gallica 144 (1): 119128.

Hack, H., Bleiholder, H., Buhr, L. Meier, U., Schnock-Fricke, U., Weber, E. and Witzenberger, A. (1992) - Einheitliche Codierung der phanologishcen Entwicklungsstadien mono- und dikotyler Pflanzen- Erweiterte BBCH-Skala, Allgemein-Nachrichtenbl. Deut. Pflanzenschutzd 44: 265-270.

Hack, H., Gall, H., Klemke, T.H., Klose, R., Meier, U., Stauss, R. and Witzen-Berger, A. (1993) - Phanologische Entwicklungsstadien der Kartoffel (Solanum tuberosum L.). Codierung un Beschreibung nach der erweiterten BBCH-Skala mit Abbildungen. Nachrichtenbl. Deut. Pflanzenschutzd 45: 11-19.

Hatzilazarou, S. P., Syros, T. D., Yupsasnis, T. A., Bosabalidis, A. M. and Economou, A. S. (2006) - Peroxidases, lignin and anatomy during in vitro and ex vitro rooting of gardenia (*Gardenia jasminoides* Ellis) microshoots. Journal of Plant Physiology 163: 827-836.

Heyman, J., Kumpf, R. P., and De Veylder, L. (2014) - A quiescent pathway to plant longevity. Trends in Cell Biology 24: 443-448.

Kerr, I. D., and Bennett, M. J. (2007) - New insight into the biochemical mechanisms regulating auxin transport in plants. Biochemical Journal 401:613-622.

Klein, J. D., Cohen, S., and Hebbe, Y. (2000) - Seasonal variation in rooting ability of myrtle (Myrtus communis L.) cuttings. Scientia Horticulturae 83: 71-76.

Koukourikou-Petridou, M. A., and Bangerth, P. (1997) - Effect of changing the endogenous concentration of auxins and cytokinins and the production of ethylene in pea stem cuttings on adventitious root formation. Plant Growth Regulation 22: 101-108.

Lancashire, P.D., Bleiholder, H., Langeluddecke, P., Stauss, R., Van den Boom, T., Weber, E., and Witzen-Berger, A. (1991) - An uniform decimal code for growth stages of crops and weeds. Annals of Applied Biology 119: 561-601.

Large, E.C. (1954) - Growth stages in cereals. Illustrations of the Feekes scale.

Plant Pathology 3: 128-129.

Larrinaga, A.R. (2010) - Rabbits, (*Oryctolagus cuniculus*) select small seeds when feeding on the fruits of *Corema album*. Ecological Research 25: 245-9.

Li, S-W., Xue, L., Xu, S., Feng, H., and Na, L. (2009) - Mediators, Genes and Signaling in Adventitious Rooting. The Botanical Review 75 (2): 230-247.

Macdonald, B. (1993) - Practical Woody Plant Propagation For Nursery Growers. Timber Press, Portland, Oregon: 219-610.

Magnitskiy, S., Ligarreto, G.M., and Lancheros, H.O. (2011) - Rooting of two types of cuttings of fruit crops Vaccinium floribundum Kunth and Disterigma alaternoides (Kunth) Niedenzu (Ericaceae). Agronomia Colombiana. 29 (2): 361-371.

Marangon, M.A., and Biasi, L.A. (2013) - Blueberry cuttings in the seasons with indolbutyric acid and substrate heating. Pesq. Agropec. Bras, Brasilia. 48 (1): 25-32.

Martine C.T., Lubertazzi D., and Dubrul A. (2005) - The biology of Corema conradii: Natural history, reproduction, and observations of a post-fire seedling recruitment. Northeastern Naturalist 12: 267286.

Martinelli, T., Andrzejewska, J., Salis, M. and Sulas, L. (2014) - Phenological growth stages of *Silybum marianum* according to the extended BBCH scale. Annals of Applied Biology 166 (1): 53-66.

Martinez-Calvo, J., Badenes, M.L., LLacer, G., Bleiholder, H., Hack, H., and Meier, U. (1999) - Phenological growth stages of loquat tree (*Eribotrya japonica* (Thunb.) Lindl.). Annals of Applied Biology 134: 353-357.

McKechnie, I.M., Burton, P.J., and Massicote, H.B. (2012)- Propagation of Vaccinium membranaceum and V. myrtilloides by seeds, hardwood stem, and rhizome cutting methods. Native Plants Journal.13 (3): 223-235.

Meier, U. (2001). BBCH-scale: refer to 'Growth Stages of Mono- and

Dicotyledonous Plants', BBCH Monograph, 2 edition, Federal Biological Research Center for Agriculture and Forestry (https://www.politicheagricole.it/flex/AppData/WebLive/Agrometeo/ MIEPFY800/BBCHengl2001.pdf).

Meier, U., Bleiholder, H., Buhr, L., Feller, C., Hack, H., Hess, M., Lancashire, P.D., Schnock, U., Stauss, R., Van den Boom, T., Weber, E., and Zwerger, P. (2008) - The BBCH system to coding the phonological growth stages of the plants - history and publications -. Journal Fur Kulturpflanzen 61 (2): 41-52.

Meier, U., Graf, H. Hack, H., Hess, M., Kennel, W., Klose, R., Mappes, D., Seipp, D., Stauss, R., Streif,J., and Van den Boom, T. (1994) - Phenologische Entwick-lungsstadien des Kernobstes (*Malus domestica* Borkh. und *Pyrus communis* L.), des Steinobstes (Prunus-Arten), der Johannisbeere (Ribes-Arten) und der Erdberee(*Fragaria x ananassa* Duch.). NachichtenbLDeut. Pflanzenschutzd 46: 141-153.

Melgarejo, P., Martinez-Valero, R., Guillamon, J.M., Miró, M., and Amoros, A. (1996) - Phenological stages of the pomegranate tree (*Punica granatum* L.). Annals of Applied Biology 130: 135-140.

Mesén, F., Newton, A. C., and Leakey, R. R. B. (1996) - Vegetative propagation of Cordia alliodora (Ruiz & Pavon) Oken: the effects of IBA concentration, propagation medium and cutting origin. Forest Ecology and Management 92: 45-54.

Mihaljevic, S., and Salopek-Sondi, B. (2012) - Alanine conjugate of indole-3-butyric acid improves rooting of highbush blueberries. Plant Soil Environ. 58 (5): 236-241.

Moubayidin, L., Mambro, R. D., and Sabatini, S. (2009) - Cytokinin-auxin crosstalk. Trends in Plant Science 14 (10): 557-562.

Moubayidin, L., Perilli, S., Dello Ioio, R., Mambro, R. D., Constantino, P., and Sabatini, S. (2010) - The rate of cell differentiation controls the Arabidopsis root meristem growth phase. Current Biology 20 (12): 1138-1143.

Negi, S., Sukumar, P., Liu, X., Cohen, J. D., and Muday, G. K. (2010) - Genetic dissection of the role of ethylene in regulating auxindependent lateral and adventitious root formation in tomato. The Plant Journal 61 (1): 3-15.

Nicolas J. I. L., Acosta, M., and Sanchez-Bravo, J. (2004) - Role of basipetal auxin transport and lateral auxin movement in rooting and growth of etiolated lupin hypocotyls. Physiologia Plantarum 121: 294-304.

Oliveira, P.B. and Dale, A. (2012) - Corema album (L.) D. Don, the white crowberry - a new crop. Journal of Berry Research. 2 (3): 123-133.

Ortega-Martinez, O., Pernas, M., Carol, R. J., and Dolan, L. (2007) - Ethylene modulates stem cell division in the Arabidopsis thaliana root. Science 317 (5837): 507-510.

Osmont, K. S., Sibout, R., and Hardtke, C. S. (2007) - Hidden branches: developments in root system architecture. Annual Review of Plant Biology 58: 93-113.Valdes B., Talavera S., and Fernandez-Galiano E. (1987) - Flora Vacular de Andalucia Occidental, Ketres Editoria SA, Barcelona, Spain, 21: 485.

Overvoorde, P., Fukaki, H., and Beeckam, T. (2010) - Auxin Control of Root Development. Cold Spring Harbor Prespectives in Biology: doi: 10.1101/cshperspect.a001537

Puri, S., and Thompson, F. B. (2002) - Relationship of water to adventitous rooting in stem cuttings of Populus species. Agroforestry Systems 58: 1-9.

Sabanti, S., Bels, D., Wolkenfelt, H., Murfett, J., Guilfoyle, T., Malamy, J., Benfey, P., Leyser, O., Bechtold, N., Weisbeek, P., and Scheres, B. (1999) - An auxin-dependent distal organizer of pattern and polarity in the Arabidopsis root. Cell 99 (5): 463-472.

Santos, C., Tavares, L. R., Pontes, V., Alves, P. M., McDougall, G. J., Stewart, D., and Ferreira, R. B. (2009) - Portuguese crowberry (*Corema album*), an interesting antioxidant white berry. 4th International Conference on Polyphenols and Health, Harrogate, UK, 7-11 December 2009 (Poster). P13.

Santos, M.S.S. (2013) - Effect of pre-treatments on the germination of seeds of the species *Corema album* L. (subsp. *album*). [Online] Lisbon: Instituto Superior de Agronomia. 89p. University of Lisbon. Dissertation for the degree of Master in Agricultural Engineering - Horticulture and Viticulture [Consult. 15 May. 2015] Available at https://www.repository.utl.pt/request-item?handle=10400.5/6462&bitstream-id=21963

Santos, M., Oliveira, C., Valdiviesso, T., and Oliveira, P.B. (2014)- Effects of pretreatments on *Corema album* (L.) D.Don (subsp.album) seeds' germination. Journal of Berry Research. 4 (4): 183-192.

Sanz-Cortés, F., Martinez-Calvo, J., Badenes, M.L., Bleiholder, H., Hack, H., LLacer, G., and Meier, U. (2001) - Phenological growth stages of olive trees (*Olea europaea*). Annals of Applied Biology 140 (6): 151-157.

Scherer, G. F. E. (2011) - Auxin-binding-protein1, the second auxin receptor: what is the significance of a two-receptor concept in plant signal transduction? Journal of Experimental Botany 62 (10): 33393357.

Schwambach, J., Ruedell, C. M., De Almeida, M. R., Penchel, R. M. Araùjo, E. F., and Fett-Neto, A. G. (2008) - Adventitious rooting of Eucalyptus globulus x maidenni mini-cuttings derived from ministumps grown in sand bed and intermittent flooding. New Forests 36 (3): 261-271.

Simmonds N.W. (1979) - Principles of Crop Improvement, Longman: 408.

Simon, S., and Petrasek, J. (2011) - Why plants need more than one type of auxin. Plant Science 180 (3): 454-460.

Strader, L. C., and Bartel, B. (2011) - Transport and metabolism of the endogenous auxin precursor indole-3-butyric acid. Molecular Plant 4: 477-486.

Tchoundjeu, Z., Avana, M. L., Leakey, R. R. B., Simons, A. J., Asaah, E., Duguma, B., and Bell, J. M. (2002) - Vegetative propagation of Prunus

Africana: effects of rooting medium, auxins concentrations and leaf area. Agroforestry Systems 54: 183-192.

Tutin T. G., Heywood, V. H., Burges, N. A., Moore, D. M., Valentine, D. H., Walters, S. M., and Webb D. A. (1972) - Flora Europea.

Diasporaceae to Myoporaceae, Cambridge University Press, Cambridge 3: 70.

Valdes B., Talavera S., and Fernandez-Galiano E. (1987) - Flora Vacular de Andalucia Occidental, Ketres Editoria SA, Barcelona, Spain, 21:485.

Van den Berg, C., Willemsen, V., Hendriks, G., Weisbeek, P., and Scheres, B. (1997) - Short-range control of cell differentiation in the Arabidopsis root meristem. Nature 390: 287-289.

Vanneste, S., and Firml, J. (2009) - Auxin: A Trigger for Change in Plant Development. Cell 136 (6): 1005-1016.

Vidoz, M. L., Loreti, E., Mensuali, A., Alpi, A., and Perata, P. (2010) - Hormonal interplay during adventitious root formation in flooded tomato plants. The Plant Journal 63 (4): 551-562.

Vieten, A., Sauer, M., Brewer, P. B., and Friml, J. (2007) - Molecular and cellular aspects of auxin-transport-mediated development. Trends in Plant Science 12 (4): 160-168.

Vignolo, G.K., Fischer, D.L., Araujo, V.F., Kunde, R.J., and Antunes, L.E. (2012) - Rooting of woody cuttings of three blueberry cultivars with different concentrations of AIB. Ciència Rural. 42 (5): 795-800.

Weber, E., and Bleiholder, H. (1990) - Erlauterungen zu den BBCH- Dezimal-Codes fur die Entwicklungsstadien von Mais, Raps, Faba- Bohne, Sonnenblume und Erbse - mit abbildungen. Gesunde Pflanzen 42: 308-321.

Wei, Y.Z., Zhang, H.N., Li, W.C., Xie, J.H., Wang, Y.C., Liu, L.Q., and Shi, S.Y. (2013) - Phenological growth stages of lychee (*Litchi Chinensis* Sonn.) using the extended BBCH-scale. Scientia Horticulturae 161: 273-277.

Woodward, A. W., and Bartel, B. (2005) - Auxin: regulation, action, and interaction. Annals of Botany 95 (5): 707-735.

Yang, Y., Hammes, U. Z., Taylor, C. G., Schachtman, D. P., and Nielsen, E. (2006) - High-affinity auxin transport by the AUX1 influx carrier protein. Current Biology 16: 1123-1127.

Zazimalova, E., Murphy, A. S., Yang, H., Klara, H., and Hosek, P. (2010) - Auxin transporters- why so many? Cold Spring Harbor Perspectives in Biology 2, a001552.

Zerche, S., and Druege, U. (2009) - Nitrogen content determines adventitious rooting in Euphorbia pulcherrima under adequate light independently of pre-rooting carbohydrate depletion of cuttings. Scientia Horticulturae 121: 340-347.

Zunzunegui, M., Diaz Barradas, M.C., Clavijo, A., Alvarez Cansino, L., Ain Lhout, F., and Garcia Novo, F. (2006) - Ecophysiology, growth timing and reproductive effort of three sexual forms of *Corema album* (Empetraceae). Plant Ecology 183: 35-46.

7. Annexes

7.1 Annex 1

yes
I want morebooks!

Buy your books fast and straightforward online - at one of world's fastest growing online book stores! Environmentally sound due to Print-on-Demand technologies.

Buy your books online at
www.morebooks.shop

Kaufen Sie Ihre Bücher schnell und unkompliziert online – auf einer der am schnellsten wachsenden Buchhandelsplattformen weltweit! Dank Print-On-Demand umwelt- und ressourcenschonend produziert.

Bücher schneller online kaufen
www.morebooks.shop

info@omniscriptum.com
www.omniscriptum.com

FSC
www.fsc.org

MIX
Papier aus verantwortungsvollen Quellen
Paper from responsible sources
FSC® C105338

Printed by Books on Demand GmbH, Norderstedt / Germany